Sampling Design and Statistical Methods for Environmental Biologists

SAMPLING DESIGN AND STATISTICAL METHODS FOR ENVIRONMENTAL BIOLOGISTS

ROGER H. GREEN
University of Western Ontario

A Wiley-Interscience Publication

JOHN WILEY & SONS
New York · Chichester · Brisbane · Toronto

Library of Congress Cataloging in Publication Data

Green, Roger Harrison, 1939–
    Sampling design and statistical methods for
environmental biologists.

    "A Wiley-Interscience publication."
    Bibliography: p.
    Includes index.
    1. Ecology—Statistical methods.   2.   Sampling
(Statistics) I. Title.

QH541.15.S72G73   574.5'01'82   78–24422
ISBN 0–471–03901–2

Printed in the United States of America

21 20 19 18 17 16 15 14 13 12 11

To the late LaMont Cole who started me off in
this direction and to Harold Welch who more
than anyone created the environment that
allowed these ideas to come into focus.

PREFACE

In the course of more than a decade of research, graduate student supervision, teaching, and general advising in basic ecology and applied environmental studies, I became more and more convinced of the need for a book such as this. Although my own field research has been in aquatic environments, I have had involvement with sampling design and statistical analysis in studies of many kinds. The problems encountered and the questions asked fell into a pattern that was independent of whether the species studied were mice, moose, or mollusks. How many samples should I take? How large should each sample be? I have more than one biological variable of interest and more than one environmental variable to explain them with—what should I do? What can I do about missing samples? How can I analyze presence-absence data? And so on.

It is not that the answers to such questions are absent from the literature or that there is a lack of good statistics textbooks. Even multivariate statistical methods are now adequately covered at a level appropriate to the nonstatistician. The answers are scattered through a diverse literature, however, and a great communication gap exists between the statisticians who know the principles of proper sampling design, hypothesis testing, and data analysis (though often more in theory than in practice) and the biologists who need to apply them to environmental studies. I wrote this book not to supplant the existing literature, but to organize and make it accessible. In a sense my motives were selfish. I was tired of answering the same questions, and making the same complaints about the results of studies where no one had bothered asking questions at all before heading into the field. I wanted also to gather this diverse information in a concise form for my own use.

I especially wish to thank the National Research Council of Canada for the financial support that enabled me to research and write this book. Upsala College in New Jersey provided me with workspace during the 1976–1977 academic year, and the City College of the City University of New York made it possible through an adjunct faculty appointment for me to interact with a graduate class in biology during autumn 1976. To both institutions I am grateful. Early drafts of sections of the book were critically reviewed by P. Handford, J. Koval, I. MacNeill, A. Storey, G. Vascotto, and W. Vickery. In particular I appreciate the contribution of A. N. Arnason to the review of literature on mark-recapture statistics. A most efficient typing job was done by G. McIntyre and L. McMullan. Finally, I wish to acknowledge gratefully the contribution of the staff and students of the Department of Zoology of the University of Manitoba, during the period 1968–1976, to my experience in this area and to the development of my concept of this book.

I apologize to anyone whose favorite methodology has been slighted. This is a selective rather than an exhaustive review of methods appropriate to environmental studies, and I take full responsibility for the selection, which is very much the product of my own experience and preferences. I assume that anyone who feels I have not placed adequate emphasis on particular methods will let me know.

ROGER H. GREEN

London, Ontario
February 1979

CONTENTS

Sampling Design and Statistical Methods for Environmental Biologists

One

INTRODUCTION

1.1 PURPOSE AND RELATION TO EXISTING LITERATURE

The purpose of this book is to provide biologists with a compact guide to the principles and options for sampling and statistical analysis methods in environmental studies. Powerful statistical methods now exist that are particularly appropriate in this field, but they and the computer programs necessary for their use are described in a widely scattered and often inaccessible literature. In the absence of a concise guide to such methods the environmental biologist tends to fall back on easy-to-use methods that are currently in fashion. In this book I discuss the principles for design of sampling and selection of statistical analysis methods, and review the types of variables and other specific decisions related to data collection, analysis, and presentation. I then present selected analysis methods and their assumptions, with references to computer programs and to applications of the methods in the published literature.

Arguments for biological assessment in applied environmental studies are well presented by Cairns and Dickson (1971). The type of problem typically confronting the environmental biologist concerns the effects of environmental factors on species in a biological community. A sampling program must be designed to obtain data for analysis by some statistical method, to test whether there is in fact evidence of an effect on the biota, and to describe efficiently any demonstrated effect. There are many variations within this general statement that define a particular environmental study, and these variations usually determine the best sequence of sampling design and statistical analysis.

By variations I do not mean specialized technical options such as the type of sampling gear to be used in a particular environment. Options of

1

this kind are not the subject of the present handbook, although in Section 3.7, I briefly review the estimation of variables in special areas (e.g., productivity, morphology and growth rates, mark-recapture methods) and give basic references. The form of the data from measurement of variables is an example of something that is relevant in all environments, and critical to choices of sampling design and statistical analysis methods. Data might be presence-absence, quantitative densities, rank abundances, or any of several other forms. The biological data might be of one form and the environmental data of another. Presence-absence data might be used in preference to quantitative data because of such constraints as cost, available time, or the possible importance of rare species. The decision to use presence-absence data then determines aspects of the statistical analysis. Some procedures designed for quantitative data are no longer suitable. On the other hand, the use of presence-absence data bypasses certain assumptions about the underlying sampling distribution for quantitative data.

The form of the data is only one of many important and interdependent considerations that together determine the best approach to an environmental study. Consider as examples the four following objectives:

1. To obtain warning of environmental deterioration at the site of an effluent by monitoring to detect change in species composition.
2. To determine the impact effects, if any, of existing point-source pollution by assessing the spatial pattern of species composition in the adjacent area.
3. To classify a series of habitats on the basis of their environments and biotas, for assignment to different categories of multiple use.
4. To determine whether a community has changed over time, given a long series of annual species lists.

These objectives have some similarities, but in fact represent different sampling and analysis problems. As shown in Table 1.1, the spatial patterns of the biota in two cases contain all or part of the information of interest. In one case they are a source of noise that obscures the required information, and in another they are irrelevant. The relevance of temporal change in the biota also varies, as do spatial pattern and temporal change in the environmental variables. In one of the four cases a pollutant exists by definition. In another it is anticipated, and in the other two the existence of a pollutant is not a major point of interest. It then follows that the appropriate distribution of sampling in space and time differs among the four objectives. Even if the same type of environment were involved in all four cases, the nature of each problem suggests a different approach

Table 1.1 Sampling and analysis properties for four environmental studies objectives

Objective	1 Warning of Environmental Deterioration	2 Pattern and Point-Source Pollution	3 Habitat Classification	4 Community Change from Species Lists
Spatial biological pattern	Noise	Information	Information	Not relevant
Natural environment spatial pattern	Noise	Noise	Information	Not relevant
Pollutant	To be detected	Exists	May exist along with other environmental factors	Perhaps relevant to subsequent evaluation
Distribution of samples in space	Experimental and control station, each with replication	Many stations on a grid, each with replication	In different habitats, preferably with replication	Not relevant
Temporal biological change	Information	Noise	Noise	Information
Natural environmental temporal change	Noise	Noise	Noise	Information
Distribution of samples over time	Equal intervals, to continue indefinitely	At least two different times	Preferably two different times	Long series, preferably at equal intervals

in sampling design, statistical analysis, tests of hypotheses, and presentation of results.

Biological field data of the kind usually obtained in environmental studies rarely satisfy the assumptions of classical statistical analysis. The undergraduate student interested in ecological work as a career is often introduced to statistics in a course that teaches all about linear models and additive, independent, and normally distributed errors. He or she then learns statistical methods based on such assumptions. Eventually the student must apply the knowledge to data obtained by sampling real organisms in natural environments, and is shocked to discover that such data almost never satisfy those assumptions. He may throw up his hands in frustration and proceed blindly, fleeing to nonparametric statistical methods or to species diversity indices. Preliminary sampling and careful evaluation of the data might have indicated that a simple transformation of the raw data would reduce the violations and their consequences to an acceptable level. Once the consequences of the violations have been evaluated, it may even be acceptable to proceed with no transformation but in full knowledge of the effect on tests of significance and on estimation of parameters.

1.2 BACKGROUND ASSUMED

This book is intended for practicing environmental biologists in academe, government, industry, and consulting firms. The background assumed is first courses in mathematics, statistics, and ecology at the university level, together with some experience with ecological fieldwork in a job or in graduate thesis research. The calculations for some examples in Sections 3 and 4 are in matrix notation. For a review of matrix algebra Davis (1973), Poole (1974), and Batschelet (1976) are recommended.

I anticipate the volume's use both as a handbook for guidance in the planning and execution of environmental studies and as a textbook in upper level courses. It is not intended as a statistics textbook, either introductory or advanced, or as a general quantitative ecology textbook. Where appropriate such text and reference books are cited in relation to particular topics.

1.3 ORGANIZATION AND EMPHASIS

The book is organized in three major sections: PRINCIPLES, DECISIONS, and SEQUENCES. In the first a general review is given of principles of inference, basic sampling and statistical design, hypothesis formulation, statistical analysis and hypothesis testing with ecological data, and effective presentation of results. This is followed by the use of a simple impact study example to illustrate ten principles for environmental biologists (see inside front cover). The DECISIONS section describes the spatial-by-temporal framework within which an environmental study is defined, then presents and discusses a key to the five main sequences of sampling design and statistical analysis methodology in environmental studies (see inside back cover), and finally reviews specific decision areas that are involved in an environmental study—such as choice of variables, estimation of necessary sample size and sample number, data screening prior to analysis, computer programs, and techniques for visual display of results. The five Main Sequences are discussed and illustrated by examples in the SEQUENCES section.

Examples are presented to illustrate methods throughout the book, but where good examples of a method are in the literature they are usually cited instead. The purpose here is to tie together a methodology that already exists but is widely scattered throughout many books and journals—it is not to reproduce the contents of that literature between these covers. Therefore there are many citations of methods, as well as computer programs for and examples of their application. In areas where the reader's background may be weak references to good review papers and basic texts are given. There is a large References section presented with symbols denoting coverage of particular topics by each reference. The subject material of different sections of this book is related, and many references are relevant to more than one subject. To avoid repetition there is extensive cross-citation, including citations to where it is used after each reference in the References section.

Obviously this book could be used as a cookbook guide to methods, but this is not my intention. Choice of statistical methods should be preceded by an understanding of principles and an application of them to the decision-making process. The book is organized with this in mind, and I urge that it be used in this way.

Certain environments, taxa, and types of impacts are used as examples out of proportion to their occurrence or economic importance. In some cases they are used because I am particularly familiar with them—marine and freshwater benthic communities, mercury, and mollusks, for example. More often, however, they are used because they provide the best examples for particular methods or because they are well treated in the literature. In any case all examples illustrate widely applicable principles and methods, with the exception of the brief review of special areas in Section 3.7. Symbols used in the References section denote references that relate explicitly to freshwater, marine-estuarine, or terrestrial environments.

Two

PRINCIPLES

2.1 GENERAL REVIEW

2.1.1 Introduction

In an environmental study there should be a logical flow: purpose → question → hypotheses → sampling design → statistical analysis → tests of hypotheses → interpretation and presentation of results. Proper statistical methods should be used, but the biologically defined objective should dominate and utilize the statistics, rather than the reverse. Commenting on the tyranny that mathematics can exercise over biological interpretation, the biometrician J. G. Skellam (1969) remarks: "Let us never forget that man earned his living as a biologist long before he became a mathematician"; he cautions that the very textbooks from which we learn statistics "show us how well the facts of Nature can be used to illustrate mathematical theorems rather than the other way round". Hurlbert (1975) observes that "biologists can often extract useful information from data that high-church statisticians would regard as hopeless, biology having advanced considerably before statistics emerged as its auxiliary discipline". Let us proceed with this attitude.

Some good general references on basic statistics for biological problems, with emphasis on the how and why and whether you use methods as well as their cookbook calculations, are Elliott (1977) and Sokal and Rohlf (1969, 1973). Elliott's small paperback book is generally applicable despite the words "benthic invertebrates" in the title, and must be one of the least known best buys available. If your French is passable, see Elliott and Décamps (1973) for a short review paper version of the book.

Many of the general principles that are stated and discussed below are

illustrated by applications to an example in the second part of this PRIN-
CIPLES section (Sections 2.2 and 2.3).

2.1.2 Statistical analysis and hypothesis formulation

Colquhoun (1971) comments that "Most people need all the help they can
get to prevent them from making fools of themselves by claiming that
their favorite theory is substantiated by observations that do nothing of
the sort. And the main function of that section of statistics that deals with
tests of significance is to prevent people making fools of themselves". It
is to prevent the reporting of nonsense that we apply hypothesis-testing
statistics, *not* to prove that nonsense is really sense. Unfortunately the
inadequacies of language obscure correct attitudes to probability, con-
ceptual models, and inference (Skellam 1969) as we deal with a world that
"is neither all chaos nor all order" (Skellam 1969 after Körner 1966).

Hypothesis formulation is a prerequisite to the application of statistical
tests. Statistical analysis can of course be used descriptively, as a kind
of exploratory dissecting kit, but if this is the case then test statistics (such
as t, F and X^2), standard errors, or confidence limits are usually inappropriate. All of these concepts are implicitly related to tests of significance,
and all such tests have meaning only when they are made against an *a
priori* null hypothesis (usually designated H_0) which can never be proved
correct but can be rejected with known risks of being wrong in doing so.

A null hypothesis must be falsifiable, which is to say that you cannot
in the end explain the results so that they support a particular hypothesis
regardless of what the results are. If the null hypothesis H_0 is that a given
impact (say an effluent) does not increase the abundance of a specified
indicator species, then you must reject H_0 if increased abundance is observed in the impact area. You cannot *then* say: "Come to think of it there
are some other good reasons why the abundance increased, so I won't
blame the impact after all". Willingness to rationalize results in this way
renders hypotheses unfalsifiable. There must be possible outcomes that
necessitate each of the two possible decisions: accept H_0 or reject H_0.

Formulate the null hypothesis by Occam's Razor, which in essence
says that your hypothesis should be the simplest one possible consistent
with the evidence, with the fewest possible unknowable explanatory factors. That an apparent increase in abundance of a species is no more than
would be expected from sampling error would be the simplest hypothesis
and an appropriate H_0. If it is rejected as being too improbable, then the
alternative hypothesis H_A should be the hierarchically next more complicated explanation. If care was taken to verify that no change occurred
other than the impact by effluent discharge then increase caused by impact

is an appropriate H_A. "God did it" is an explanation that is possible, but is unfalsifiable and the ultimate violation of Occam's Razor. (This H_0 and H_A notation is used throughout the book.)

Remember that "the problem must dictate the methods to be employed for its solution, not the reverse" (Redfield 1958). Do not get the cart before the horse and hunt around for cookbook statistics which then determine what hypothesis is being tested. Let your purpose generate the question, and let the null hypothesis be the simplest possible answer to that question, stated in a way that is testable and falsifiable. Then worry about sampling and statistics. Simplicity of the H_0 and H_A are desirable but sometimes impossible to obtain. For example Hinckley (1969) describes radiation-induced fluctuations in forest insect populations where the biological system is so complex that an appropriate H_0 (unimpacted population trends) would be difficult to formulate in a simple, unambiguous, and falsifiable manner. Great care and careful thought are required in the planning and execution of such studies.

The H_0 and the H_A should differ only with respect to the question being asked. In the situation described by Hinckley, for example, it would not be appropriate to test for significant fluctuations in insect abundance. This implies an H_0: "insect populations stay at static levels." They do not, of course, even in the absence of radiation. This may seem obvious, but environmental biologists are forever using statistical tests that do the equivalent of this. The significance of R^2 in 10-variable multiple regressions or of the overall chi-square in large contingency tables is not likely to provide a basis for meaningful decision-making in environmental studies. Such statistical models describe a potentially complex array of things that could be going on (as an implied H_A). An implied H_0 of "nothing whatsoever is going on" does not provide a very useful contrast. Hypothesis testing should be more like the use of a rifle than a shotgun.

When statistical analysis is used for exploratory descriptive purposes tests of H_0: "nothing is going on" may be appropriate. Multivariate procedures such as ordination and clustering are often used to reduce massive, complex data sets to "what is going on." Since such techniques can yield results that give the appearance of things going on even when applied to data simulated to be completely random, it is a wise precaution to reject H_0: "nothing is going on" first, before proceeding to carry out such descriptive analyses. Description implies that you know there is something to describe. Williams and Lance (1965) and Goodall (1966a) discuss this problem in relation to such multivariate procedures.

2.1.3 Philosophy of inference and hypothesis testing

The test of a null hypothesis results in a decision to accept or reject, based on some estimated risk of being wrong in that decision. It is usually the probability that one will reject H_0 when it is in fact true (the Type I error) that is of most concern. The largest acceptable risk of this error's occurrence is commonly set at $\alpha = 0.05$. The 5 percent level of significance is only a convention, but anyone who uses a higher probability level should be prepared for the suspicion that the level was chosen *after* determining that the results were not quite significant at $\alpha = 0.05$.

If H_0: "no biological damage resulted from an impact" is tested at $\alpha = 0.05$ then a significant result means that on the evidence there is less than a 1-in-20 chance that the observed conditions would have occurred naturally—namely, for reasons other than impact effects—and that once out of 20 or more times is an acceptably low risk of being wrong in the conclusion that there *were* impact effects. Testing at $\alpha = 0.01$ indicates that the risk of a Type I error, of concluding impact effects when there were none, must be 1-in-100 or lower to be acceptable.

However, for any given sampling and statistical analysis design, lowering the Type I error level α will increase the Type II error level β, which is the probability of concluding that H_0 is true when in fact it is not. Here one would falsely conclude that no impact-related biological damage has occurred. There is always a tradeoff, and the only way to reduce one error level without increasing the other is to improve the design. For example, one could increase the number of samples, which would permit reduction in either Type I or Type II errors or both.

2.1.4 Models

The ability to make such statements with assurance depends on more than statistical tests. There must be a proper H_0 that is expressed in terms of a statistical model. The model states some hypothesized functional relationship between one or more criterion, or dependent, variables (measures of biological impact effects perhaps) and one or more predictor, or independent, variables. The latter could be measures of impact intensity and perhaps also of environmental variation which is *not* related to impact but may contribute to variation in the criterion variable(s).

All models are abstractions and simplifications of reality. As Redfield (1958) notes, science *is* the construction of models of nature. Modeling in the narrowest systems analysis sense, as "a burgeoning subscience" practiced by environmental engineers and civil servants who lack biological intuition and mistake the mathematics of the models for reality itself, receives a scathing critique from Hedgpeth (1977) which is recommended reading for all who are in their first fascination with modeling.

In environmental studies a model that completely described the complex systems involved would have to have hundreds of simultaneous partial differential equations with time lags and hundreds of parameters (Levins 1966). Simplication is both legitimate and necessary, and Levins notes that one must choose among three possible strategies in model-building. The perfect model would have the attributes of generality, realism, and precision. In practice one of these must be sacrificed, and which it is defines the strategy. Though no models are perfect, "What really matters is not their degree of perfection, but their adequacy for prescribed purposes" (Skellam 1969). Generality may be sacrificed if one is interested only in predicting what will happen at a single location. Nonlinear models based on assumed instantaneous responses (e.g., the logistic model) sacrifice realism but may be adequate for many purposes. There may be no need for high precision if the biological change to be detected is a large one. Skellam emphasizes that while much thought is usually given to errors in estimation, and to hypothesis testing, too little is given to "a much more serious source of error and deception, the defects of the model itself." Environmental biologists are among the worst offenders in this regard; they are far too willing to collect massive sets of data, with no clear hypothesis model in mind, and then throw them into the nearest computer package multiple regression program.

2.1.5   Design of sampling

Good general references are Steel and Torrie (1960), Cochran (1963), and Elliott (1977). A discussion of pollution survey design is in Cairns and Dickson (1971).

It is critical that at some level the sampling be random or most statistical tests will be invalid. The assumption of independence of errors is the only one in most statistical methods for which violation is both serious and impossible to cure after the data have been collected. Truly random allocation of samples is the necessary and sufficient safeguard against this violation.

The sampling design should reflect the H_0 model as closely as possible. If H_0: "an impact has resulted in no biological damage" contrasts with an H_A that it has, then the study should be designed so that as many as possible of the environmental variables unrelated to impact are controlled, perhaps by choosing control and impact areas that are similar except for the fact that one will be impacted and the other will not. Then the sampling design itself will hold constant extraneous environmental variables which may be correlated with the biological criterion variable(s). Design for such control of extraneous variables is considered by Gruenberger (1964) to be the most important in his list of 13 attributes of good science.

Careful thought must be given to the choice of criterion and predictor variables. Are they really measures of what you want to know (impact effects and degree of impact, say)? Is the cost and time necessary for their measurement reasonable? How many variables need be measured? Crovello (1970) emphasizes that all this forms a multistage decision process. For example, in the impact study situation the first decision would be what system to use. Suppose that the benthic community is chosen, which will constrain choices of both criterion and predictor variables. If organisms are used as criterion variables, the actual taxa and the taxonomic level to which they will be identified (genus? species?) must be decided. And so on. Crovello points out that "since most workers usually have some idea about the system and what characters might be more closely related to the purpose and more reliable, the preliminary screening of characters is done more or less informally in the researcher's head."

Samples of the population are in the end collected by some device or method that will rarely be 100 percent efficient or unbiased. You must ask yourself what population you are really sampling. The one you think you are? Suppose that fish density is a criterion variable. To what extent will catch per standard otter trawl tow represent density? What about bottom fish? What about fast swimmers and avoidance? Perhaps the population you are *really* sampling is the activity-abundance of those species and age classes of fish that are susceptible to otter trawls. If so, are you still measuring a criterion variable adequate for your purpose? If small mammal density in a field is a criterion variable, to what extent will snap-trap lines provide a measure of it? Are you not probably sampling a population from an area larger than the field, in that foreign animals will tend to move into vacated territories and then be caught? Will you not be measuring activity as much as or more than density? If so, what are the implications for your study and its objectives? Such questions should always be asked, and they can only be answered by doing some preliminary sampling to evaluate your sampling method (see Section 2.3.5).

The distinction between random errors (resulting in decreased precision) and bias must be kept clearly in mind. Bias is the magnitude and direction of the tendency to measure something other than what was intended (Eisenhart 1968). "Statistical methods can only cope with random errors and in real experiments systematic errors (bias) may be quite as important as random ones. No amount of statistics will reveal whether the pipette used throughout an experiment was wrongly calibrated" (Colquhoun 1971). Nor will statistics reveal that on the average a sampling method collects only 70 percent of the population that is really there.

2.1.6 Choice of statistical analysis

It should again be emphasized that the choice of an appropriate statistical analysis should flow logically from the purpose, the hypothesis, and the sampling design. The hypothesis model should determine the statistical model. As indicated in British Columbia Lands, Forests and Water Resources (1974), "No set procedures can be given because each situation is to some extent unique. The analysis used depends on the information required and on how the data are collected. It is therefore imperative that the purpose of the study and the expected methods of analysis are known before the samples are collected."

Any statistical analysis method assumes certain things about the data. There are nonparametric (or distribution-free) methods, but in general a tradeoff occurs in that methods with fewer assumptions perform less powerful tests of the hypothesis. More detailed discussion of assumptions can be found in Section 2.1.7 and throughout the book in relation to particular methods.

Besides appropriateness to the hypothesis model, what criteria should be applied to the choice of statistical analysis? Should it be "easy to use"? This approach has led many environmental biologists to the use of diversity indices (see Section 3.5.2). I would argue, rather, that the analysis should be the most efficient one appropriate to the hypothesis model and that the results then should be explained and interpreted in nontechnical terms to lay audiences.

An efficient statistical analysis method will be as conservative, powerful, and robust as possible. If it is conservative, it will have a low probability α of making the Type I error—concluding that there were biological effects of the impact when in fact there were none. If it is powerful, it will have a low probability β of making the Type II error—concluding there were no biological effects when in fact there were. If it is robust, the stipulated error levels will not be seriously affected by the kinds of data commonly encountered in environmental studies.

A final cautionary note regarding tests of significance and the present availability of computers to do them: "Perhaps the greatest danger . . . is the temptation it places before the unwary investigator. One can generate tests of significance in such volume that the uninitiated may ignore the 95 percent lacking in significance at the .05 level, and write the papers about the significant 5 percent. With the rapidity of the computer one may fool oneself all the more quickly" (Jones 1964).

2.1.7 Problems with ecological data

The kinds of data commonly encountered in environmental studies rarely satisfy the assumptions of the statistical methods taught in undergraduate courses. Missing observations for some variables at certain times and places are common, and the data set to be used for hypothesis testing may be a mixture of different kinds of variables: binary, ranks, quantitative, and so on. Few statistics texts, even those that are biologically oriented, devote much space to consideration of these problems and the options available when confronted with them.

Two basic references on the assumptions of classical statistical methods, and the consequences of their violation, are Eisenhart (1947) and Cochran (1947). They are still well worth reading. Glass et al (1972) have written an excellent review paper on this subject, emphasizing published results from application of statistical methods to data simulated with specified violations of assumptions. In other words their emphasis is on consequences of violations in practice rather than in theory. This paper is little known to biologists, although it should be, for the conclusions are reassuring (see Sections 2.3.3 and 2.3.9).

Assumptions of independent and normal error distributions, homogeneity of error variation among groups or along regression lines, and additivity of effects are usual for analysis of variance and related statistical methods. The need for truly random sampling to ensure independence of errors has already been mentioned. Violations of the remaining assumptions commonly occur together, rooted in a single cause. For example, samples of abundances of organisms tend to produce skewed distributions, a tendency toward multiplicative rather than additive effects, and heterogeneity of variances caused by functional dependence of the variance on the mean (Cassie 1962, Elliott 1977). Of these the most serious violation is heterogeneity of variances. Methods for testing against H_0: "homogeneity of variances", are illustrated in Section 2.3.9. It is also a wise precaution to examine scatter plots and histograms of residuals (errors) for evidence of gross violation of these assumptions, and this is easily done. Techniques are discussed by Cleveland and Kleiner (1975) and Daniel and Wood (1971). The Statistical Analysis System (SAS) is one of several computer packages that allows such plots to be easily produced (Service 1972, Barr et al 1976).

The robustness of statistical methods in the face of violations of assumptions may be evaluated by simulating data that satisfy the H_0 model but have the undesirable properties of the real data (the same kinds of nonnormal error distributions, heterogeneous variances, etc.) and by applying the proposed statistical method to them. Several examples of such evaluations are reviewed by Glass et al (1972). Hurlbert (1969) and Maelzer (1970) use this approach to evaluate statistics that supposedly describe population density-dependence and interspecific association, respectively. A different analysis method for species co-occurrences is evaluated by Crawford and Wishart (1967). Fager (1972) provides an example of simulated sampling from various random and nonrandom spatial distributions typical of organisms.

It is of course possible to use biological field data for the evaluation of statistical methods (e.g., Lambert and Williams 1966) but "Such an approach is unavoidably circular, because the results of a given analysis can only be compared with a preconceived idea of what is in the data, or with what some other analysis method suggests is in the data. The difficulty is that the true properties of the data are not known" (Green 1977). Simulation of data for this purpose is not difficult. Small data sets can be simulated on programmable pocket calculators such as the Texas Instruments SR-52 (see Sections 2.3.9 and 4.1). It is possible to simulate large sets and then directly input them to statistical analysis procedures using SAS. Programs in the Cornell Ecology Programs Series (Gauch 1976) are designed for simulation of multispecies samples from entire communities distributed over environmental gradients. The use of such a program to evaluate spatial pattern, sample unit size, shape and orientation, and statistical analysis methods is described by La France (1972). Hilborn (1973) describes a control system for Fortran simulation programming. Examples of statistical methods applied to simulated data are presented at various places throughout this book.

Simulation may also be used to validly test hypotheses when data violate the assumptions of the methods. With a new generation of students who are at ease with computers, this approach to hypothesis testing will probably eventually replace cookbook statistics. A typical procedure would be as follows. Calculate the standard tests of significance as if the assumptions did hold, but realize that they do not estimate correct probability levels. Then simulate a number of sets of data satisfying H_0 but having the same properties that violate the assumptions. If the test statistic value for the observed data exceeds the test statistic values for, say, 95 percent of the simulated H_0 data sets, then H_0 may be rejected at $\alpha = 0.05$ (see Section 2.3.9). Maile (1972) uses this approach for testing whether two groups differ on several variables. If a systems analysis model has been developed to the point where it adequately describes the previous history of a natural system, then simulation may be used to test for change by providing a prediction based on known factors against which the observed events can be compared. Two examples are Parker (1968) and Herricks and Cairns (1974). A good general review paper on simulation is that by Raeside (1976).

Robustness of decisions reached by statistical analysis will be increased by the use of several analysis methods based on different assumptions. With this in mind Crovello (1970) argues for the use of as many methods as possible at any given stage of analysis. I would emphasize very strongly that this is *not* an argument for trying different analysis procedures until one of them gives you the result you want. Rather, it is the opposite: "if these models, despite their different assumptions, lead to *similar* [my

italics] results we have what we can call a robust theorem which is relatively free of the details of the model. Hence our truth is the intersection
of independent lies'' (Levins 1966). This approach is particularly useful
when carrying out exploratory analyses on multivariate data sets where
hypotheses are vaguely defined. For example H_0: "any structure is continuous rather than discontinuous" contrasted with H_A: "there are natural
groups, or clusters, of the n samples defined by the p variables" suggests
cluster analysis procedures. If several clustering procedures, based on
different algorithms related to different definitions of a group, all produce
the same clusters of samples, the conclusion that there *are* real groups
of some kind is a robust one.

Finally, one may seek robustness of analysis methods, and of the conclusions based on them, in nonparametric methods. Two good general
references are Siegel (1956), and Mosteller and Rourke (1973). One should
however consider whether nonparametric methods are necessary. Glass
et al (1972) describe "a largely unnecessary hegira to nonparametric statistics during the 1950s," remarking that "the assumptions of most mathematical models are always false to a greater or lesser extent. The relevant
question is not whether . . . assumptions are met exactly, but rather
whether the plausible violations of the assumptions have serious consequences on the validity of probability statements.'' They conclude that
the flight to nonparametrics was largely unnecessary. There is no hard
and fast rule. If the most efficient possible method appropriate to the
hypothesis model turns out to be one based on ranks, then so be it. For
example one of the best ordination procedures, called nonmetric multidimensional scaling, is based on ranks (see Section 3.4.2) and for many
kinds of ecological data it may be the most efficient ordination model that
is at all realistic in its assumptions. On the other hand, the biologist who
resorts to nonparametric versions of ANOVA because of slight nonnormality of within-group error distributions is being unnecessarily picky.
The price paid for a slight increase in robustness is a loss of power in the
test and a loss of information in descriptive presentation of results. Mosteller and Rourke themselves emphasize that a rigid cookbook approach
should not be followed, and that it is often best to design your own test
(see their Chapter 16) or to use simulation as described previously.

Nonparametric methods are used in examples throughout this book
where they are appropriate.

2.1.8. Presentation of results

The attributes of scientific communication that distinguish between the
scientist and the crackpot are discussed by Gruenberger (1964). Some

attributes have to do with quality of the research itself, while others deal with the manner in which the results are presented.

Present the methods you used in sufficient detail that they can be repeated by someone else and your conclusions verified. Avoid unnecessary details, however. No one cares to know the dimensions of the concrete block through which the effluent pipe runs or that it was difficult sampling from your small boat on windy days.

If possible your results should be stated in such a way that they can be used for prediction of future events. Correct prediction, say of certain biological effects given certain kinds and levels of impact, will increase your credibility.

A scientifically acceptable style of writing should be used. Such guides as Council of Biology Editors (1972) consider organization and writing style; American Mathematical Society (1962) describes proper mathematical and statistical notation, symbols, formats for equations, and so on.

One should tend more toward humility than arrogance when reviewing the results of others. If you think someone is wrong it is proper to say so and to say why, but not to belittle him or her either explicitly or implicitly. Declarations about how important your results are should be avoided, as should exclamation points implying that you just said a good one that might not be noticed. Remember that Watson and Crick (1953) first described the DNA double helix structure simply as having "novel features which are of considerable biological interest." Avoid paranoia, the "This is the true way but the establishment is against it" attitude.

Avoid statistics compulsion. Although this book emphasizes the use of statistics in environmental studies, nothing is worse than inappropriate or unnecessary use of statistics. Graphical or tabular data formats that show obvious differences can stand on their own. Rules for making clear and concise tables and graphs are discussed in Lewis and Taylor's (1967) Chapter on "Analysis in Ecology." Crovello's (1970) review paper has a section on graphic presentations. Excellent examples are in a paper by Rosenberg (1973), which describes the response of a benthic community in a Swedish fjord to pulp mill effluent. Figures 2.1 to 2.4 use a variety of techniques to effectively demonstrate spatial and temporal effects of pollution in terms of species composition, species number, and abundances. Figure 2.5, from a paper by Hughes and Thomas (1971), effectively shows differences among species in their depth distributions, which are of such magnitude that statistical tests would only be overkill.

The proper balance lies somewhere between the attitude that if you need statistics your results aren't any good, and the attitude of the compulsive referee who demanded a statistical test from a biologist to prove

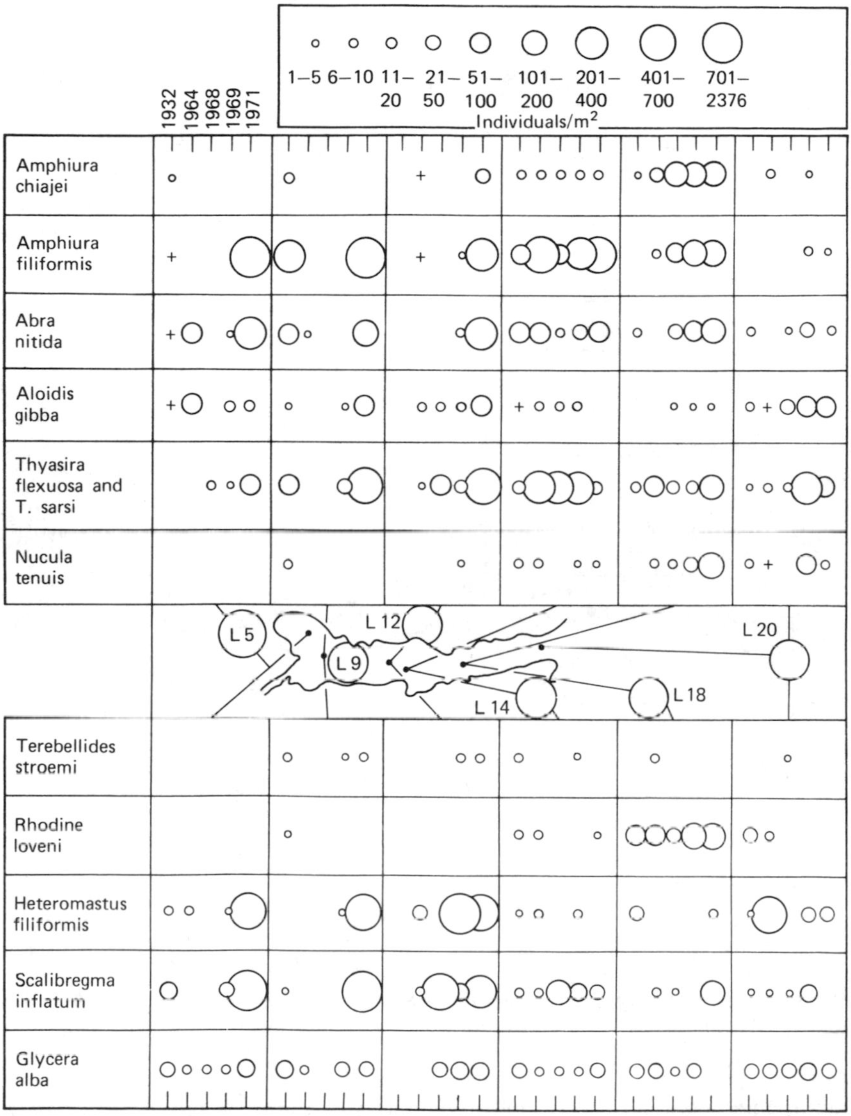

FIGURE 2.1 Distribution and density of some benthic species in a Swedish fjord (Saltkällefjord) between 1932 and 1971. An outfall from a sulphite pulp mill entered the fjord just below station L5 in the early 1960s until the mill operation ceased in 1966. Reproduced with permission from Figure 9 of Rosenberg (1973).

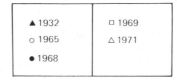

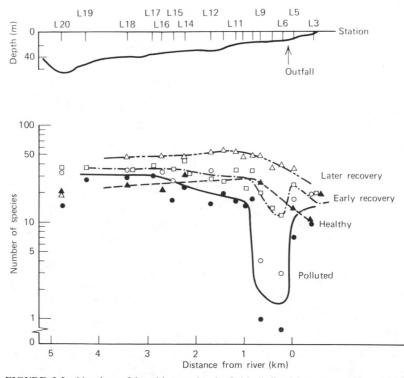

FIGURE 2.2 Number of benthic species in Saltkällefjord between 1932 and 1971. See legend for Figure 2.1. Reproduced with permission from Figure 5 of Rosenberg (1973).

significance when all 1000 nematodes chose chamber A rather than chamber B given a choice and equal access. Emphasize "minimal recourse to statistics" (Hurlbert 1975). When possible emphasize use of statistics for choosing a sample number and design that will allow convincing mean differences to be demonstrated without additional statistics, rather than to prove slight, unconvincing differences. "Ideally, . . . significance tests should never appear in print, having been used, if at all, in the preliminary stages . . ." (Colquhoun 1971).

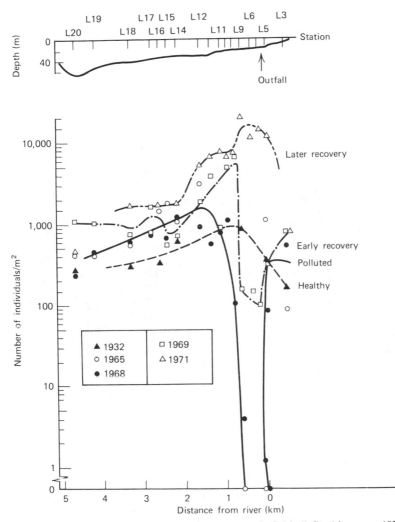

FIGURE 2.3 Number of individuals per square meter in Saltkällefjord between 1932 and 1971. See legend for Figure 2.1. Reproduced with permission from Figure 6 of Rosenberg (1973).

When the uncertainties of final results are expressed in the written report they should be unambiguous, but not redundant. For example vertical bars above and below mean values on graphs, or "$a \pm b$" expressions, are ambiguous unless it is clearly stated what they represent. They could be standard deviations, standard errors, or confidence limits of some kind. On the other hand, it would be redundant to present test statistics

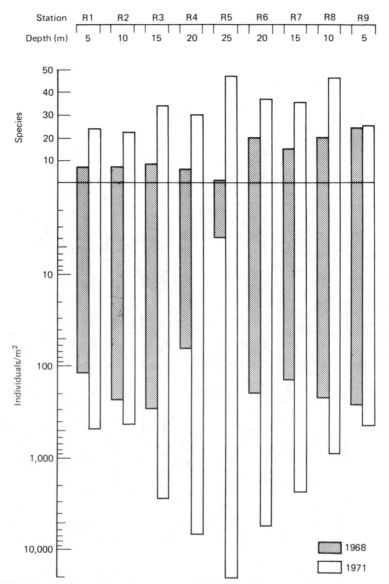

FIGURE 2.4 Number of benthic species and individuals in Saltkällefjord at stations along a transect just below the sulphite pulp mill outfall. Faunal recovery between 1968 and 1971 is shown. See legend for Figure 2.1. Reproduced with permission from Figure 4 of Rosenberg (1973).

(such as X^2 or F-values), their degrees of freedom, the probability level, *and* a symbol such as **. A biologist recently presented a correlation coefficient along with the number of samples on which it was based, and was told by a referee to provide enough information so that significance could be judged. He had, of course, done so. Eisenhart (1968) summarizes the appropriate presentation of the uncertainties of final results according to four cases:

1. Bias and imprecision are both negligible—give the result to the significant number of digits and state what that number is.
2. Bias is not negligible—state that the imprecision is negligible and state what the accuracy is.
3. Neither bias nor imprecision is negligible—separately state the bounds on each, clearly acknowledging that there is bias as well as imprecision.
4. Imprecision is not negligible—give the standard error and state explicitly that there is no bias.

2.1.9 Interpretation of results

A possible source of error is the computer program itself. The program may not be working properly because it has not been debugged after being set up for use on that system or because the program blows up with inappropriate input data. Some matrix operations critical to statistical analyses will not work where there are too many samples, too many variables, variables coded outside certain ranges, too high a ratio of variables to samples, or for other reasons. Unfortunately, it may not always be obvious that this is happening. The program may run and yield results with severe rounding errors. Certain steps in statistical analyses, such as matrix root and vector solutions in multivariate analyses, should be done in double (16-place) rather than single (7-place) precision. Rounding error can be a problem in single precision programs even when the data are well coded (see Section 4.4).

Another possible source of error is the confused nomenclature that is used in computer program outputs and in manuals for statistically oriented desk or pocket calculators. Results can easily be misinterpreted. Statistical methods such as covariance analysis and discriminant analysis tend to be used for different purposes in, say, medicine and agriculture, and programs based on the same mathematics but designed for different purposes often yield output that looks totally different. Significance tests can easily be interpreted as being tests of hypotheses that are not, in fact, being tested. In programs for analysis of variance the F-tests are usually based on the

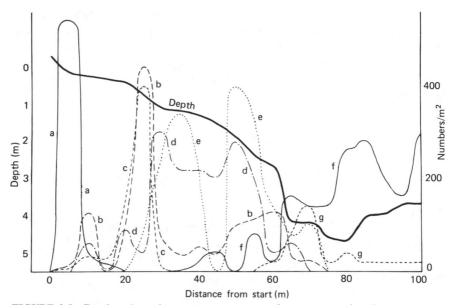

FIGURE 2.5 Depth and numbers per square meter of common species along a transect perpendicular to shore in a Prince Edward Island estuary. a: *Mya arenaria*; b: *Mytilus edulis* (X2); c: *Nassarius obsoletus*; d: *Neopanope texana* (X20); e: *Crepidula fornicata* (X20); f: *Yoldia limatula* (X20); g: *Nephthys incisa* (X20). Reproduced with permission of the authors from Figure 15 of Hughes and Thomas (1971).

among-replicate variance estimate as the denominator, which is not always appropriate (e.g., see Section 4.1). Even something as simple as a variance cannot be assumed to be correct. Often the population variance formula (dividing by the sample number n instead of by $n - 1$) is used in calculator programs without that fact being clearly stated. Sometimes the sample variance formula *is* used but the calculator key is marked with the symbol σ^2, which should denote the population variance.

There is a simple, necessary, and sufficient precaution for these calculator- and computer-derived errors: *always* try out previously analyzed data in a new program or package of statistical programs to check the precision, accuracy, and interpretations of the results. Only then proceed to use it for analysis of your biological field data. Examples from widely used textbooks by reputable authors can often serve the purpose.

We all acknowledge that a statistically significant relationship is not proof of causality, but in practice it is easy to slip into the assumption of causality with no independent proof of it. Correlation of variables in time series is often abused in this way, and we can all laugh at correlations between ozone levels and library circulation (Huntington 1945) and at tongue-in-cheek biological clocks in the unicorn (Cole 1957a). In class I present data that "prove" that the annual cycle of abundance of polychaete worms in Barnstable Harbor, Massachusetts, is strongly correlated with the salinity of Moreton Bay, Queensland. The fact that everything we study is on the same earth confounds a great many variables in environmental studies. Very sophisticated statistical methods have been developed for time series analysis (see Section 4.2), and with such methods the danger of misinterpretation due to confounding of variables is perhaps even greater. Certainly, time series analysis is a necessary and valuable tool for some problems (Hamon and Harmon 1963, Maslow and Keith 1971, Sinha et al 1972), but interpretation of the results always requires great caution.

Confounding of variables is often characteristic of spatial as well as temporal relationships. For example, in streams environmental variables such as current velocity, depth, sediment particle size, organic content, and sorting coefficient are strongly intercorrelated (e.g., De Marche 1976). Descriptive field sampling alone cannot show whether a statistical relationship between a species distribution and, say, sediment type has direct causal basis. An experimental approach, allowing independent manipulation of environmental variables that are strongly correlated in nature, becomes necessary. Transects across environmental gradients are particularly vulnerable to the confounded variable problem, for example estuary-to-open ocean transects (e.g., Bird 1970) where salinity, depth, and substrate all change together. As Pielou (1974, 1975) has demonstrated, distributions of species on environmental gradients may not be explainable solely in terms of the environmental variables, even if all relevant ones are considered. The effect of competition may be to produce species zones or even more complex patterns on environmental gradients. In such a situation a statistical model relating species abundances to any environmental gradient variables would be completely misleading. Leonard (1970) and Fenchel (1975) provide examples, from the terrestrial and the marine environment respectively, of intrinsic biological variation which in a descriptive study could easily be misattributed to environmental variation. The difficulty of actually separating social and environmental causes of species associations is emphasized by Cole (1957).

When morphological criterion variables are used with environmental predictor variables, there can be confounding with morphological clines such as that described by Cvancara (1963) for a freshwater clam. One might determine that decreased size was a predictable response by a particular species to a given pollutant, and then fall into the trap of applying the size-versus-pollutant statistical model at a distant location where the unpolluted size is different.

Confounding of the sampling method or the statistical analysis method with the predictor or criterion variables is another common problem in environmental studies. For example, benthic sampling devices such as grabs vary in efficiency when used on different substrates. Where the purpose of the study is to predict species abundances from substrate properties that are in practice confounded with the sampling efficiency, there is—to say the least—a problem in interpretation of the results. De Benedictis (1973) has shown that such statistical constructs as species diversity indices are usually correlated by definition—that is, by the formulae that define them. This would seem to represent an intrinsic confounding, which invalidates any attempts to interpret biologically the correlations among different diversity indices, and yet such interpretations continue to appear in the literature.

Finally, with presence-absence or quantitative species abundance data the apparently simple result of zero abundance, or absence, can pose serious interpretation problems. If a species is present, in any abundance, one at least knows that it can live there (Hutchinson 1968, Cairns 1974) but absence of that species from two samples need not indicate environmental similarity of the samples. In one sample the level of some environmental variable may be too extreme on the high side and in the other too extreme on the low side, or different environmental variables may be responsible for the absence in the two cases, or the species may be absent for historical zoogeographic rather than proximate environmental reasons (Green 1971a). Because absence has many causes, interpretation is particularly difficult when the area sampled includes many depauperate locations (Crossman et al 1973). Cairns (1974) particularly emphasizes the dangers of methods that use absence as an indicator in environmental studies.

2.2 THE EXAMPLE

2.2.1 The nature of the problem

The problem is one commonly encountered in applied environmental studies. An effluent will be discharged into a river at a known time in the future and thereafter. We are told to design and execute a field study, and to carry out necessary statistical analyses of the data obtained therefrom, to determine whether the effluent discharge results in biological change.

2.2.2 The data

The advantage of using data simulated to have known properties, for evaluation of statistical methods, was discussed in Section 2.1.7. The simulated data used here are included in the larger data set used in Section 4.1 to illustrate a multivariate statistical approach to this problem. The procedure by which the total data set was simulated is discussed in some detail there. Here I only state the properties of the data as they apply to this example.

The species whose abundance, as determined by appropriate sampling, is the biological criterion variable is lognormally distributed with standard deviation σ equal to 50 percent of the mean abundance (see Section 2.3.9). Where the species is unaffected by the effluent the mean abundance is 50 organisms per square meter. In the area just below the effluent discharge the mean abundance is reduced by 75 percent.

2.2.3 Presentation of the example

This example is presented to illustrate 10 basic principles of sampling design and statistical analysis in environmental studies, which should be followed step-by-step in the development and execution of any study. With a few exceptions, each is organized as follows: (a) the principle, (b) justification, (c) references to people who followed the principle and those who did not follow it, and (d) illustration by the example. The example is regarded throughout as based on a real problem, in which the properties of the data are *not* known *a priori*. Only when a result is obtained, a decision made, and a judgment required is reference made to the true properties of the data.

2.3 TEN PRINCIPLES

2.3.1 Be able to state concisely to someone else what question you are asking. Your results will be as coherent and as comprehensible as your initial conception of the problem.

Begin by stating your objective in commonsense terms (Does an effluent cause biological damage?). Then rephrase the question as a statement (The effluent causes biological damage.) Increase the precision of the statement by including within it information about the criterion and predictor variables (The abundance of benthic species Y as measured by sampling device C is reduced in an area just below the effluent discharge in comparison with an area unaffected by the effluent.) Let this be the hypothesis H_A (see Section 2.1.2). Now formulate the null hypothesis H_0 as the hierarchically next simpler hypothesis, differing from H_A only in the question of interest (The abundance of benthic species Y as measured by sampling device C is *not* reduced in an area just below the effluent discharge in comparison with an area that is unaffected by the effluent *but is otherwise the same*.)

The exact formulation of the H_0 and H_A above represent a number of important decisions that have been made regarding choice of system, criterion and predictor variables, and sampling method (see Section 2.1.5). These decisions, and some approaches to them, are discussed in Section 3.6. Species Y may be of value for its own sake (a commercially valuable shellfish, say), previous studies may have shown it to be sensitive to the pollutant involved (a good indicator—see Section 2.3.8), or it can be shown to have high information content regarding the species abundance composition of the entire community. Sampling device C may have been described as the most efficient and reliable for sampling that species in that environment. Here the hypotheses H_0 and H_A imply that the sampling design itself will control the two levels (effluent, no effluent) of the environmental predictor variable.

The next step is formulation of the sampling design and statistical model to allow an efficient test of the hypothesis H_0 against the alternative H_A (Section 2.1.4). The sections that follow deal with various aspects of this. You should pause now, however, and check that you have clearly formulated what you want to do. Go to someone who knows the community involved (so you can speak the same biological language), but not the particular study you are planning, and explain what you want to do by concisely stating the objective and the hypotheses to be tested. If he or she understands it, then so do you.

If you need expert statistical advice the time to seek it is now—after you have formulated what you want to do and why, and before you have limited (perhaps destroyed) your statistical options by proceeding to sample and collect some or all of your data. Environmental biologists often say that statisticians are hard to talk to, and not of much use. Certainly there are statisticians who hate to dirty their theories by applying them to real problems. Often, however, the fault lies with the biologist, who is not sure what he or she wants from the statistician other than some kind of salvage operation. If you have delayed seeking expert advice until you can only ask "What can I do with my data?", you richly deserve, at that point, any answer you get.

Spight (1976) refers to "the arbitrary design and undirected confusion of contemporary environmental studies" which lack tests of "well-stated hypotheses." Lambert (1972) remarks that "Too often vast quantities of raw data are collected without enough thought in advance as to how, and to what end, and at what cost, the material can be subsequently handled." If you remember nothing else, remember to heed this first principle. Then you will at least know what you are doing and for what your statistical analyses are intended. The likelihood that you will reach meaningful conclusions and make clearcut decisions will be much improved.

2.3.2 Take replicate samples within each combination of time, location, and any other controlled variable. Differences among can only be demonstrated by comparison to differences within.

Hypotheses are usually tested by determining whether the ratio of the variation caused by the hypothesized effect (difference in mean abundance of species Y between the effluent-affected area and the unaffected area) to the error variation (that within each area) is larger than would be expected if the null hypothesis H_0 (as stated in Section 2.3.1) were true. If certain assumptions are satisfied (see Sections 2.1.7 and 2.3.9) so that the estimates of variation are estimates of true *variances* (and covariances), we have the F-statistic, or some multivariate analogue of it. This concept is clearly developed and illustrated in Chapter 7 of Sokal and Rohlf (1973). The argument for this second principle is simply that an estimate of the magnitude of variation within as well as of variation among is required for most tests of hypotheses, and the former requires replication. In terms of our example, we will ensure that there is replication within each area each time it is sampled.

The following examples illustrate the importance of this principle. A graduate student began research on the effect of depth and temperature on metabolic rate of a zooplankton species throughout the year. Without seeking statistical advice he spent his first year going out at monthly intervals and suspending single bottles containing constant densities of the organism for fixed times at various depths. After a year's work he took his data, consisting of single metabolic rate estimates for each depth-by-time of year combination, to a statistician. "What can I do with my data?", he asked. "Nothing, except to graph and discuss it," was the reply. Result: a wasted year of work that had to be repeated.

A consulting firm studying effects of heated effluent discharges on fish in a large American river did seek statistical advice and developed an appropriate sampling design involving standard trawl tows, in replicate, at different times and locations. Halfway through the study a cost-cutting campaign in the company resulted in an order to save on plastic buckets used aboard ship, by throwing the fish from replicate tows into one bucket. This practice of pooling replicates is not uncommon, and usually demonstrates that the worker has heard somewhere that replication is good but does not really understand why. Of course, obtaining a mean value for observations derived from the replicate samples is pooling too. The same mean will be obtained if the replicates are *physically* pooled before identification and tabulation, but all basis for estimating variation within will be lost. "Compositing" of replicate samples *can* be appropriate in some circumstances, allowing "one to sample a larger area with minor increases in costs" where "Selection of a primary sample is cheap while actual determination of density is costly and time-consuming" (Rohde 1976, extending the results of Brown and Fisher 1972). Rohde's example is of "sampling large bodies of water in order to estimate the density of phytoplankton." Such an approach should be considered to be a special case calling for careful planning of sampling design and analysis.

2.3.3   Take an equal number of randomly allocated replicate samples for each combination of controlled variables. Putting samples in "representative" or "typical" places is not random sampling.

In Section 2.1.5, it was emphasized that independence of errors is the only assumption whose violation is impossible to cure after the data have been collected, and that truly random sampling will prevent that violation. In their review of "Consequences of failure to meet assumptions . . ." Glass et al (1972) conclude that correlated errors can have more serious consequences on the validity of tests of significance than all other violations. If the errors are positively correlated, the test will tend to be more liberal than the nominal level (say, $\alpha = 0.05$); if negatively correlated, more conservative.

It requires effort to ensure that sampling is random. Left to his or her own impulse the investigator will probably space out the samples by avoiding locations close to previous samples, or will avoid locations that are difficult to sample or obviously inhospitable to the organism. If rocks are avoided during core sampling on a beach to estimate crab density, the density is for that part of the beach which is not rock. When sampling a mosaic of habitats, some of which are more likely to contain the organism than others, this characteristic should be included in the sampling design by randomly sampling within habitat-type strata (see Section 2.3.7). When randomly sampling, randomness should be ensured by choosing locations from tables of random numbers. Single numbers will establish positions on transects, pairs of numbers on two-dimensional surfaces, and so on. If the random numbers table was left behind, the sequence of numbers formed by the last two digits of phone numbers in the nearest directory will do well enough.

Many statistical analyses can be carried out where there is unequal sample number, for example, one-way ANOVA or simple linear regression. However, many analyses that can be particularly appropriate for powerful overall tests either require equal sample number or more difficult calculations for unequal sample number (Steel and Torrie 1960). See also Section 3.5.3 for further discussion. This is true of factorial ANOVA designs (as used in Sections 2.3.9 and 4.1). The interaction components of variation in factorial analyses are difficult to interpret unambiguously when sample number differs among treatment combinations. Another argument for equal sample number is that violation of the ANOVA assumption of homogeneity of within-group variances appears to have serious consequences on significance tests only when the sample numbers of the groups differ and are correlated with the group variances (Glass et al 1972).

A good example of an environmental study where samples consistent in number and in mode of collection are allocated to each combination of the location and time of the controlled variables is that by De Marche (1976). For our example we conclude that an equal number, at least two, of replicate samples should be collected from each area at each time of sampling, and that the exact locations of the samples within the areas should be defined by coordinates drawn from a random numbers table. Determination of the appropriate number of replicate samples is discussed in Section 2.3.8.

2.3.4 To test whether a condition has an effect, collect samples both where the condition is present and where the condition is absent but all else is the same. An effect can only be demonstrated by comparison with a control.

Design of sampling so that control and impact areas correspond as closely as possible to the hypotheses H_0 and H_A was discussed in Section 2.1.5. In a review of work on secondary effects of pesticides in aquatic ecosystems Hurlbert (1975) argues that the majority of studies have been weak in design and statistical methodology, which has in turn weakened the conclusions. He strongly emphasizes the lack of and the need for proper controls.

Evidence for impact effects on the biological community should be based on changes in the impact area that did not occur in the control area. The need for controls in both space and time in optimal impact study designs is discussed further in Section 3.2. Controls are particularly necessary in toxicology or uptake experiments where the unnatural conditions under which the organisms are held can cause mortality or other effects even at zero dosage level. Also, the administration of the active compound may require procedures that, in addition to the effects of the compound itself, stress the organisms. For example, Koblynski and Livingston (1975) describe studies of uptake of the insecticide mirex by a fish, where the mirex had to be mixed with hexane in order to administer it. The hexane, without mirex, was added to the control aquarium so that only the effect of interest (the mirex) would differ.

In our example the need for a spatial control, as an area unaffected by the effluent but otherwise the same as the area just below the effluent discharge, has been stated (Section 2.3.1). We will sample both areas before the effluent discharge begins, which will provide a temporal control, as well as sampling both areas after the effluent has been discharging for some time. If the density of species Y decreases in the impact area but not in the control area, we will reject H_0 and accept H_A. The sampling design is shown diagrammatically in Figure 2.6. Three randomly allocated

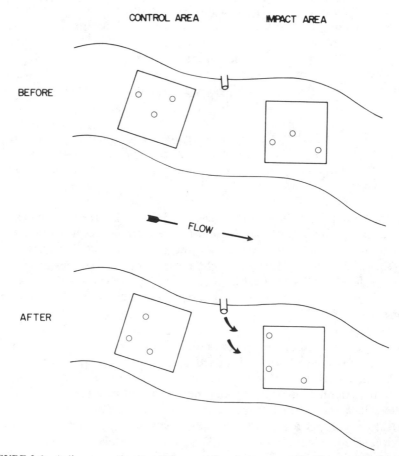

FIGURE 2.6   A diagrammatic view of the sampling design as used in the example. See text for discussion.

replicate samples per area-by-time combination are shown, for purposes of illustration.

2.3.5 Carry out some preliminary sampling to provide a basis for evaluation of sampling design and statistical analysis options. Those who skip this step because they do not have enough time usually end up losing time.

The importance of preliminary sampling is probably the most underemphasized principle related to field studies. There is no substitute for it, and it is seldom advisable to skip it. There is always insufficient time, and psychologically it is difficult to commit oneself to spending time on sampling which will probably not generate data for the final results. The situation, however, is similar to that of a sculptor who skips the preliminary model in clay or wax only at his or her own peril. The time spent will probably be time saved later, though we all find time saved in the hand to be more attractive than time saved in the bush.

The results of the preliminary sampling *can* be included in the final results which are analyzed and interpreted. It is usually *not* legitimate to simply sample, determine how many more samples are needed, and then combine the additional samples with those already taken. Some modifications to the statistical analysis procedures are necessary. So-called two-stage designs allow this to be done properly (see Sampford 1962 and Cochran 1963).

The argument for preliminary sampling is simply that there is no other way to check out things that can potentially be serious problems in an environmental study: the efficiency of the sampling device, the size of the sample unit, the number of samples required to obtain the desired precision of estimates, and the possible presence of large-scale spatial pattern that would make stratification desirable. An example of the consequences of failure to do preliminary sampling is my own experience of learning it the hard way.

As a beginning graduate student I decided on a project that required year-round sampling of the benthos at a depth of 120 m in Cayuga Lake. The purpose was to obtain unbiased samples of the population of an amphipod in order to determine its life history. I decided to use a modified (by myself) version of a small dredge, and as soon as possible I rushed out and began sampling. Sorting and processing of the samples was a laborious task, so several months of sampling had been done by the time I had sorted, counted, measured, and tabulated the results. Another few months passed before I realized that something was wrong—I was not obtaining juveniles in any numbers. Then I did what I should have done in the first place: sample on the same day at the same location both with

my device and with standard benthic grabs. The latter sampled the juveniles, which obviously (*now* obviously) were deeper in the sediment.
My device did not, and had not. Result: A wasted field season. So much
for saving time.

In addition to Cochran (1963) the manual by Elliott (1977) provides a
good review, with examples, of the use of preliminary samples in planning
the sampling program. In toxicological studies it is usually desirable to
conduct preliminary runs using very wide differences in the level of toxicant, in order to establish the critical range for determination of the LD_{50}
(see also Section 3.8). An example is Davis and Hidu (1969). In our effluent
impact study example we conduct preliminary sampling in both areas, to
verify that they are similar before impact both environmentally and in the
abundance of species Y, to assess the operation of the sampling device
(including use of SCUBA to observe it in operation) and to determine the
optimal sampling design. If patchiness of the environment within areas
is suspected, the preliminary sampling should be clustered (or nested),
with random samples taken at randomly selected locations within each
area. It will then be possible to determine whether such clustering, and
what allocation of samples in it, provides most efficient sampling for the
actual study (Cochran 1963). See Section 2.3.7 for further discussion and
Section 4.1 for an example of the use of this design.

*2.3.6   Verify that your sampling device or method is sampling the
population you think you are sampling, and with equal and adequate
efficiency over the entire range of sampling conditions to be
encountered. Variation in efficiency of sampling from area to area
biases among-area comparisons.*

The problem of inefficiency and bias in the sampling method was introduced in Section 2.1.5, and the effect on misinterpretation of results was
discussed in Section 2.1.9. No area of environmental studies is immune
to this problem. In mammal studies there are problems with trap-prone
and trap-shy animals, movement into vacated habitats, and saturation of
the trap-line. In fisheries studies gill net catches depend on activity, size
of fish in relation to mesh size, and saturation of the net during a set.
Trawls and towed nets actually sample a volume of water much smaller
than the product of the mouth area times the distance towed, because of
the pressure buildup which is a function of the size of the net cone, the
mesh size, the towing speed, and the condition of the net. Fishes and even
larger zooplankton tend to avoid the net opening (Fleminger and Clutter
1965, McGowan and Fraundorf 1966). The dependence of sampling efficiency of dredges and grabs on substrate type was mentioned in Section

2.1.9. An extreme is reached with stream benthos, where active animals can be found down to 70 cm in exactly those coarse substrates that are difficult to sample quantitatively to any great depth (Coleman and Hynes 1970, Williams and Hynes 1974). When specialized methods are used, density estimates are obtained that are 10 times greater than those commonly reported from standard surface sampling methods (Hynes et al 1976). This ceases to be an esoteric example when one considers the voluminous literature of the last decade that purports to describe the environmental health of streams, using samples obtained by the same methods that can underestimate by an order of magnitude the community that is really there. Not only is the community as a whole underestimated, but the bias varies among taxonomic groups so that percentage composition estimates are badly biased as well (Carle 1976). See below and Section 3.5.2 for further discussion on this point. This same problem in relation to biased estimation of percentage species composition of zooplankton by sampling with towed nets is discussed by McGowan and Fraundorf (1966), who found that smaller nets yielded lower diversity estimates.

In the sampling of many environments, aquatic or terrestrial, the organisms of interest must be extracted in some manner from the sample as originally collected. In benthic sampling, for example, the choice of screen mesh size for sieving is critical (Reish 1959, Bloom et al 1972). Preliminary sampling should include determination of the largest mesh size which retains all of the individuals of the species being sampled. For example, trial sieving of samples of the intertidal clam *Gemma gemma* showed that newly released juveniles passed through a 297 μ but were retained on a 250 μ screen (Green and Hobson 1970). Flotation methods are often used to separate animals from lighter plant material and heavier inorganic material (Lackey and May 1971, Flannagan 1973a). Again, there are the problems both that any inefficiency will be bias, on the underestimation side, and that the bias varies among taxonomic groups. For example, Flannagan shows that such groups as mollusks and caddis flies in particular are underestimated by flotation methods. There can be bias against certain size individuals of the same species as well (Sellmer 1956).

The preceding comments should not be taken as a criticism of the specific methods mentioned. The best sampling and extraction methods in all kinds of environmental studies are inefficient and biased to some degree. This sixth principle states that you should spend the time and effort necessary to determine what that bias is and what the consequences are likely to be in relation to the environmental range you will be sampling for the purposes of your study. Bias, if you are aware of it, *may* not be a problem. If a consistent sampling methodology is applied to a control and an impact area and the level of the criterion variable differs between the areas, the conclusion about an effect of the impact may be valid regardless of whether the levels described depend on the sampling methodology. The levels described *must not*, however, depend on a difference between the areas which is unrelated to the impact, such as differing substrate type influencing the efficiency of the sampling device.

Gage (1972) provides a good example of checking the efficiency of operation of a bottom dredge at a variety of depths and on different substrates by sending a diver down to watch the dredge in operation. Carle (1976) considers the biases associated with standard methods for estimation of stream benthic populations, and especially estimation of community diversity measures that are a function of proportional abundances of different taxonomic groups. He concludes that the biases are likely to have been serious in much published work but shows how estimates can be corrected for bias by the use of a new sampler in combination with a removal sampling technique. A computer program is provided for calculation of the removal estimates.

Species diversity estimation provides an excellent example of when biased sampling is and when it is not a problem. The same comments would apply to biased estimation of other criterion variables as well. There is little doubt that estimates of true community diversity from samples are always biased. This is so even when bias specific to the sampling device itself is not a problem (Pielou 1969, p. 231). However, the Brillouin formula for the diversity H of a completely censused population can be used as a measure of the diversity of the *sample itself*, treated *as* a population. This is a legitimate approach if, and only if, one ceases to pretend that one is estimating a conceptualized true community diversity based on samples from it. Can anything useful be concluded if samples from control and impact areas have differing species diversities that cannot be used as estimates of diversity of the community from which the samples were collected? Leaving aside the problem of biological interpretation of species diversity used as a criterion variable (see Sections 2.3.1 and 3.5.2) the answer is yes *if* one can assume that the biases are the same for both areas. This is a risky assumption. For this reason the use of species diversity as a criterion variable is perhaps most justified where the population diversity H is calculated for communities colonizing artificial substrates of identical composition and manner of exposure to colonizing propagules. However, Crossman and Cairns (1974) have found that the populations colonizing artificial substrates are often so unrepresentative of the natural community that conclusions about impact effects on the latter are difficult to make. This is a matter of judgment. The health of canaries can be a good indicator of the safety of coal mines even though canaries are not natural inhabitants of coal mines. The fact that they tend to succumb to coal gas just before humans do allows practical conclusions to be made from the death of the canary.

An example of failure to verify that a benthic sampling device was providing unbiased estimates of a population was given in Section 2.3.5. An example involving bias in an extraction method is that of a graduate student who was studying organisms living between sand grains on a beach. He used an extraction method that had been described in the literature as 85 percent efficient. That is, the population would be consistently underestimated by 15 percent. *After* two years of sampling, extracting, and compiling data he checked the efficiency of the extraction method by comparing it with laborious total counts, sand grain by sand grain. The method on which he had based two years of work was approximately 35, not 85, percent efficient and the errors among replicate extraction runs were greater than the sample means.

As indicated in Section 2.3.5 we verify that in our study the sampling device is operating in an unbiased manner and with adequate efficiency over the range of environments being sampled in both areas, and that processing methods (such as sieving of samples) are retaining all of the individuals of the criterion variable species.

2.3.7   If the area to be sampled has a large-scale environmental pattern, break the area up into relatively homogeneous subareas and allocate samples to each in proportion to the size of the subarea. If it is an estimate of total abundance over the entire area that is desired, make the allocation proportional to the number of organisms in the subarea.

In Section 2.3.3 it was emphasized that allocation of samples should be either completely random or randomly allocated within strata that are chosen so that each stratum includes a portion of the total area which is a relatively homogeneous environmental type. For example, assume that the null hypothesis is that a microtine rodent species does not differ in abundance between an impacted area and a control area which are similar except for impact (Section 2.3.4). If both areas are a mosaic of meadow and forest and the species concerned is, say, a meadow vole, random allocation of sampling effort within the two areas would represent a very inefficient design. The error variation (among trap locations within areas) would be inflated by forest-versus-meadow differences in abundance, and the ratio of variation among (between the control and impact areas) to variation within would be thereby reduced, and so would the power of the test against H_0 (see Section 2.1.6). Whether this relationship between the density of the criterion variable species and the environmental mosaic is known from previous work by others, or from your own preliminary sampling, the appropriate sampling design would be a stratified one where meadow and forest represent each of the two strata. The variation between strata can thereby be removed from the analysis, so that only the variation within strata goes into the error term. The general references given at the beginning of Section 2.1.5 are applicable here. Estimation of gain in efficiency from use of a stratified as compared with a completely random design is described, and examples given, in Steel and Torrie (1960) and Cochran (1963).

When sources of variation are hierarchically related, or when the environment is known to be spatially patchy but not on a sufficiently large scale to define strata, another approach may be appropriate (Section 2.3.5). Nested or subsampling designs are particularly useful for preliminary sampling because one can with relatively little sampling effort estimate the amount of variation that is derived from each hierarchical level, and then determine the optimal allocation of sampling effort to each level for the planned study (Cochran 1963). For example if a survey for spruce budworm eggs is planned, where H_0 is that the mean density of eggs per needle has not exceeded a critical level that previous work has indicated is the outbreak threshold density, then sources of variation in the variable "eggs per needle" might be among trees, among vertical layers within trees, among branches within vertical layers within . . . . The hierarchical relationship of the sources of variation is apparent. The information obtained from well-designed preliminary sampling might indicate that by far the most important source of variation is among trees and that the optimum allocation of every, say, 100 samples (needles) would be to randomly select 25 trees in the study area and then to randomly select 4 needles from each tree without regard to vertical height or branches. Where spatial patchiness on different scales may represent important sources of variation, a hierarchical nesting of sampling spatially may be appropriate. Green and Hobson (1970) used this approach in preliminary sampling prior to a study on the clam *Gemma gemma* and found that random allocation of samples within 10 m square areas was optimal.

If determination of density variation over the study area is of primary interest, allocation of samples to strata should be proportional to stratum area. If, however, the purpose is to estimate total abundance over the entire study area, allocation should be proportional to the number of organisms in the stratum, as estimated by the preliminary sampling. Allocation within strata generally should be random. However, see Lambert (1972) for an example of systematic, or evenly spaced, sampling within subareas, allocated in proportion to the environmental heterogeneity of the subareas. Also see Cochran (1963, Ch. 8) for a discussion and review of systematic sampling in which he concludes that sometimes it can improve precision, especially where variation is nearest to continuous. Sampling on a spatial grid pattern is often convenient when the problem is assessment of impact from spatial pattern (see Section 4.4). The results of ordination or cluster analyses can then be plotted on the map of the area for effective visual display (see Section 3.11). If randomly allocated replicate samples are collected at each location on a grid, the environmental biologist will have the best of both statistical worlds.

Estimation of a commercial clam population by use of a stratified random design is described by Russell (1972). Grassle et al (1975) give an example of a combination stratified-nested sampling design where the sampling is by photographs from a deep sea submersible at depths ranging from 500 to 1800 m. Three sources of variation in abundance of epifaunal organisms are determined: among depths (which are the strata), among photographs at the same depth, and within photographs. Sometimes nested sampling at different spatial scales can be built into the sampling device itself, as described by Hamilton et al (1970) for benthos and by Tonolli and Tonolli (1958) for plankton. Estimation of variation at different spatial scales (i.e., estimation of "clump size") from preliminary sampling of terrestrial plant communities is extensively treated by Greig-Smith (1964), and related methods are presented by Zahl (1974) and Mead (1974).

In impact studies the affected area may have a definable spatial pattern. After preliminary sampling has defined the boundaries of subareas that are relatively homogeneous with regard to degree of impact, a stratified design would be appropriate for studies of spatial variation of biological or environmental variables over the entire area. The patterns of impact of effluent plumes, such as the pulp and paper mill effluent studied by Leslie (1977), Minns (1977), and others, often fall into this category. In the same study Johnson (1977) shows how a dye can be used as a marker to track the plume and to estimate dilution over the study area. See Section 3.8 for further discussion.

An example is in Green (1968), for an intertidal population study. The study as reported is based on a stratified random sampling design where the 10 strata were tidal level zones. This design greatly reduced the among replicate sample (error) variation in estimated organism abundance because of the strong pattern of density variation with tidal level. Unfortunately when I first began the study I felt that I didn't have time for preliminary sampling because I knew I was limited to a one-year study. So I randomly sampled along a transect across the intertidal for the first two months before I realized how low the efficiency of my estimates was, which I could have learned in one day of proper preliminary sampling.

Instead I lost several months and had to arrange for sampling to be continued for that length of time after I had left in order to complete the one-year cycle. Had I done the preliminary sampling I would have lost one sampling day.

For our effluent impact example we assume that preliminary sampling has shown no important gain in efficiency from stratified or nested sampling within areas-by-times combinations. Therefore we will base the analysis on a completely random design, where the number of samples to be allocated to each area at each time must be determined (Section 2.3.8). However, in the multivariate example in Section 4.1 a patchy environment suggests a nested design, which is used there as the basis for statistical analysis and hypothesis testing.

2.3.8   Verify that your sample unit size is appropriate to the sizes, densities, and spatial distributions of the organisms you are sampling. Then estimate the number of replicate samples required to obtain the precision you want.

Choice of a sample unit size depends on several factors. Logistic considerations may be one. Some large benthic grabs require a large vessel with a power winch. Some seining and netting operations for fish and for birds require several persons to carry them out. Cost of sample collection is a related factor, and where the cost tends to be proportional to the number of samples rather than to the total area (or volume) sampled the use of fewer large samples rather than many small ones might be necessary. High cost of sample processing could be a similar factor. Cochran (1963) discusses choice of sampling unit size and shape in his Chapter 9, and works out an example where information about different sources of cost are brought into the decision process. Another factor influencing choice of sample unit size is the size and shape of the organism being sampled. The ratio of the area or volume of the organism to the area or volume of the sample unit should be negligibly small—0.05 or less, as a rule of thumb. When sampling mobile organisms where avoidance is a problem (see Section 2.3.6) the same rule might be followed using the area or volume of avoidance movement produced by that particular sampling method.

The spatial distribution of the organism is an important factor influencing choice of sample unit size. The common failure of ecological data to satisfy the assumptions of many statistical methods was discussed in Section 2.1.7, and the likely consequences and possible remedies are considered in Section 2.3.9. However, at this stage, when sampling methods and design can still be modified based on information from preliminary sampling, the problem can be greatly reduced. Nonrandomness, usually of the contagious or aggregated type where the variance exceeds and is a function of the mean, depends very much on sample unit size (Greig-Smith 1964, Buzas 1970) and shape (La France 1972). Where spatial pattern is hierarchical, with patches on one scale distributed in some manner within patches on a larger scale, the question "What is a good sample size?" may have more than one answer. Sample unit sizes that coincide with any scale of patch size should be avoided, since these will maximize the among replicate sample, or error, variation. Use of nested quadrat designs or of partitioned sampling devices for preliminary sampling, described in Section 2.3.7 in relation to choice of sampling design, can also be used to determine the most efficient size of the primary sample unit.

Where a simple aggregated spatial distribution model applies, which is not one of hierarchical scales of patchiness (at least over the ranges of sample unit sizes that are practical), more general principles can be stated. The best sample unit size generally is the smallest possible one, when sampling a given total area (or volume) for an organism that has an aggregated spatial distribution. However, the qualifications are important, because in practice organism size, avoidance behavior, cost per sample, and other factors *are* important. Elliott (1977) reviews this "the smaller the better" concept extensively, as well as the related subject of the use of indices of nonrandomness to measure degree of aggregation.

In my experience most environmental biologists intuitively feel that "the larger the better" is the best rule for sample unit size because they will then obtain more per sample. It is a useful exercise to plot out simulated random and aggregated distributions on graph paper and then sample them with sample units of different sizes, proving to yourself that sample unit size makes no difference with random distributions but that smaller size results in increased precision of estimates with aggregated distributions. See Section 3.6 for an example and for approaches to choice of sample unit size which are useful in particular situations.

The best sample number is the largest sample number, keeping in mind that no sample number will compensate for poor sampling design. Precision of estimates of mean values will increase with sample number but there is a law of diminishing returns. The standard error of the mean decreases in proportion to the square root of the sample number so that an increase in sample number from 4 to 9 reduces the standard error (and the width of confidence limits) by a third. To achieve another reduction by a third a sample number of 21 would be required, and to achieve another, 46.

The most powerful argument for a large sample number is that most statistical analyses tend to be robust in the face of violations of assumptions *if* they are based on a large number of error degrees of freedom (see also

Section 2.3.9). This is primarily because of the consequences of the central limit theorem which states that the mean of n samples taken from almost any normal or nonnormal parent distribution will itself have a distribution that approaches normality as n increases. It is "the most important theorem in all of probability and statistics from both the theoretical and applied points of view" (Lee 1971). (There are exceptions—some ratios do not behave well when used as variables in statistical analyses because they do not go to normal in this manner when sampled. This is a good reason to avoid using derived variables which are ratios of other variables; see Section 3.5.3.) Barrett and Goldsmith (1976) randomly sample several data sets using different numbers of samples to investigate the question of when n is sufficiently large for the confidence interval on the mean to be valid regardless of the nonnormality of the data. They find that for slightly skewed data any sample number larger than 2 is adequate. For a moderately skewed lemming weight distribution at least 10 were required, and for a highly skewed redwood age distribution at least 40.

If it was not possible to conduct preliminary sampling and a number must be pulled out of a hat, three replicates per treatment combination is a good round number. Remember that it is the overall error degrees of freedom that are important. If mean density is estimated from simple random sampling then the error degrees of freedom is $n - 1$. If sampling is at $i = 1$, t times at each of $j = 1$, s locations with r replicate samples per location-by-time combination, it is $rts - 1 - (ts - 1) = ts\,(r - 1)$, which is $r - 1$ multiplied by the number of sets of r replicate samples in space and time (ts). In our example, with $t = 2$ times and $s = 2$ areas, it is $(2)\,(2)\,(r - 1) = 4r - 4$. Therefore, the error degrees of freedom corresponding to replications $r = 3$, 4, and 5 would be, respectively, 8, 12, and 16. Designs can be complicated past belief, but in the end it is the degrees of freedom associated with the denominator (the H_0 variation) in the test of significance that count. One of the advantages of a carefully thought out sampling and statistical design is that relatively few replicates per treatment combination can go a long way in producing robust and powerful overall statistical tests.

Estimation of sample number required to achieve a desired precision is treated extensively by Sampford (1962), Cochran (1963), Greig-Smith (1964), Elliott (1977), and Elliott and Décamps (1973). Approaches to estimating sample number in particular situations, and some rules of thumb, are considered further in Section 3.8. Two examples are presented here, the second of which is for our effluent impact problem.

Say that you wish to estimate the mean density of a species population within plus or minus 10 percent, with a 1-in-20 chance of being wrong in the conclusion that you have done so. If simple random sampling over the area is the design, and the sampled distribution is approximately normal, then let preliminary sampling yield the following values (of number per square meter) for 10 samples:

$$291 \quad 420 \quad 140 \quad 223 \quad 219 \quad 195 \quad 248 \quad 251 \quad 163 \quad 292.$$

The estimates of the sample mean and standard deviation are $\overline{X} = 244.2$ and $s = 79.16$ respectively. Confidence limits on the mean are given by

$$\overline{X} \pm t_{1-(1/2)\alpha} \frac{s}{\sqrt{n}}$$

or in this case by

$$244.2 \pm t_{1-(1/2)\alpha} \frac{79.16}{\sqrt{n}}.$$

A 1-in-20 chance of being wrong means .95 confidence limits, so that $\alpha = 0.05$, and .95 confidence limits of plus or minus 10 percent of the mean gives

$$24.42 = t_{.975} \frac{79.16}{\sqrt{n}}$$

where the only unknowns are the t-statistic and the sample number n. If t is known, we can solve for n, but t is a function of its degrees of freedom which are $n - 1$. However, for large sample sizes t is a very weak function of n and $t_{.975}$ is approximately 2. Therefore,

$$24.42 \approx (2) \frac{79.16}{\sqrt{n}} \quad \text{and} \quad n \approx 42.$$

For $n - 1 = 41$ degrees of freedom $t_{.975} = 2.02$ and if we solve for n again, using this t value, we find little difference. If the estimated n had been small, several interative solutions for n and t would have been necessary. To have some margin of safety we would probably round up to the nearest multiple of 10 and take $n = 50$ samples to estimate the species population density.

The 10 preliminary sampling values were in fact taken from a normal distribution with mean $\mu = 300$ and standard deviation $\sigma = 100$. When I did take 50 additional random samples from that distribution I estimated $\overline{X} = 303.0$, $s = 95.22$, and .95 confidence limits of 303 ± 27, which is 303 ± 9.0 percent. So it worked. A sequential procedure for sampling until a desired precision is obtained is illustrated in Section 3.8.

In our effluent impact example we will be sampling in each of two areas at each of two times (Section 2.3.4), and the areas are similar to each other in the before-impact situation that would apply during preliminary sampling. It is assumed that the area means and the within-area variation of the species abundance criterion variable are the same for both areas (see Section 2.3.9). Therefore, preliminary random samples from both areas may be pooled, yielding the following 10 values (of number per square meter):

$$56 \quad 78 \quad 74 \quad 46 \quad 30 \quad 59 \quad 30 \quad 36 \quad 80 \quad 57.$$

It is likely that it will be appropriate to logarithmically transform such species abundances, as $Z = \log{(X + 1)}$, and we will proceed on this basis without justifying it here (see Section 2.3.9).

What do we want to test? If the abundance of species Y decreases in the impact area but not in the control area, we will reject H_0 and accept H_A (Section 2.3.4). That is all very well, but what magnitude of decrease do we want to be able to detect and what error risks are we willing to accept? The impact might cause a real reduction of $\frac{1}{10}$ percent, but the number of samples required to detect it would be prohibitive in any field sampling situation. The criterion variable should have been chosen so that a realistically detectable change would occur if H_A rather than H_0 were correct (Section 2.3.1), in other words, chosen so that H_0 is falsifiable (Section 2.1.2). In this case let us say that we want to be able to detect a decrease of 50 percent in the abundance of species Y in the impact area, in contrast to zero decrease in the control area. Given a decrease of this magnitude we are willing to accept a 0.05 risk of making the Type I error (Section 2.1.3)—concluding that there is a decrease in the impact area and none in the control area when in fact there is no real impact area decrease. The efficiency of the statistical analysis method used to test H_0 and the actual magnitude of any decrease will determine the power of the test, which is the probability $1 - \beta$ of *not* making the Type II error (concluding that the abundance of species Y in the impact area stayed the same when in fact it *did* decrease).

For the statistical analysis we will use a 2×2 (before-after times and control-impact areas) factorial analysis of variance, and the test of H_0 will be the test against H_0: "no interaction" (see Section 2.3.9). Estimation of necessary sample replication, within each of the four areas-by-times combinations, to satisfy the above-stated conditions is as follows:

1. Logarithmically transform the preliminary sample data as described above. Calculate $\bar{Z}_1$ (the subscript indicates "at time 1") and s_Z^2 which are 3.9608 and 0.1341 respectively. A $\bar{Z}_1 = 3.9608$ value corresponds to $\bar{X}_1 = e^{Z_1} - 1 = 51.5$, and a decrease of 50 percent will reduce that

to $\overline{X}_2 = 25.75$ which corresponds to $\overline{Z}_2 = 3.2865$. This is a change $\triangle \overline{Z}$ of -0.6743 in the impact area, which we want to be able to detect.

2. In a 2×2 factorial ANOVA with r replicates per area-by-time combination, where SS = sum of squared deviations, df = degrees of freedom, B and A = before and after, and C and I = control and impact, the interaction is

$$F\,(1, 4\,(r - 1)\,\text{df}) = (SS_{\text{int}} \div 1\,\text{df})\,/(\,SS_{\text{err}} \div 4\,(r - 1)\,\text{df})$$

$$= SS_{\text{int}}\,/\,\text{error mean square}$$

$$= SS_{\text{int}}\,/\,0.1341.$$

$$SS_{\text{int}} = [(\Sigma_{AC} + \Sigma_{BI}) - (\Sigma_{BC} + \Sigma_{AI})]^2 \div 4r$$

$$= [(3.9608r + 3.9608r) - (3.9608r$$

$$+ 3.2865r)]^2 \div 4r.$$

Therefore,

$$F\,(1, 4\,(r - 1)\,\text{df}) = [(0.6743r)^2 \div 4r]/0.1341$$

$$= 0.8581r.$$

3. For values of

$r =$	2	3	4	5	6,
$F_{.95}\,(1, 4\,(r - 1)\,\text{df}) =$	7.71	5.32	4.75	4.49	4.35,
$0.8481r =$	1.70	2.54	3.39	4.24	5.09.

Therefore $r = 6$ replicate samples should be randomly allocated per areas-by-times combination, for a total of $4 \times 6 = 24$ samples in both areas at both times.

2.3.9 Test your data to determine whether the error variation is homogeneous, normally distributed, and independent of the mean. If it is not, as will be the case for most field data, then (a) appropriately transform the data, (b) use a distribution-free (nonparametric) procedure, (c) use an appropriate sequential sampling design, or (d) test against simulated H_0 data.

The criteria for choice of statistical design, the problems of violation of statistical assumptions with ecological data, the consequences of those violations, and some of the remedies such as nonparametric methods were briefly discussed in Sections 2.1.6 and 2.1.7. Strictly speaking, these decisions should have been made earlier in the sequence, based on the preliminary sampling data. Harris (1975), for example, gives as one of the conditions for reasonableness of a transformation that "The transformation must be *a priori*, that is, specified (or at least specifiable) before the actual data obtained in the study are examined." However, it is presented in this section because it is very much interrelated with the statistical analysis and tests of hypotheses.

Those who use statistical methods in environmental studies tend to fall into one of two distinct groups regarding the assumptions underlying the methods they use. Either they ignore the fact that there are assumptions at all or they are paranoid about them and rely on nonparametric methods. What is argued here is that (*a*) the assumptions of the method should be understood at the time it is chosen, (*b*) the liklihood and consequences of violation should be assessed (with the aid of data from preliminary sampling), and then (*c*) use of the method should proceed with awareness of the risks and the possible remedies. As noted in Section 2.1.7, nonparametric methods are rarely necessary (Glass et al 1972). Harris (1975) remarks that assumptions such as normality and homogeneity of within-group variation "are almost certainly *not* valid for any real set of data— and yet they are *nearly* valid for many sets of data. Moreover, the fact that a particular assumption was used in deriving a test does *not* mean that violation of that assumption invalidates the test, since the test may be quite robust under . . . violations of the assumptions used to derive it." He notes that there is much strong evidence that most univariate normal distribution-based statistical tests are "extremely robust" under such violations. "The major exceptions . . . occur for very small and unequal sample sizes [i.e., numbers of samples] and/or one-tailed tests." The desirability of large and equal numbers of samples was discussed in Section 2.3.8 and 2.3.3 respectively. Harris suggests some general guidelines. Tests on a single correlation coefficient will be valid for any unimodal X and Y population if n is larger than 10. Two-tailed tests with F and t-statistics will generally be valid, even on extremely nonnormal populations. The ratio of the largest to the smallest sample variance should not exceed 20, and the ratio of the largest to the smallest sample size should not exceed 4. The error degrees of freedom should be 10 or more. For discussion of the robustness of multivariate statistical methods to violations of assumptions see Section 4.1.

Even when methods are robust to violations of assumptions it is often possible to reduce the violations, including that of nonlinearity of variable relationships when using linear model statistics (such as most regression analyses), by appropriately transforming the original data. But when is a transformation appropriate, and how should you decide which one to use? The very word "transformation" is suspect, apparently synonymous with alteration or data massaging. Sokal and Rohlf (1973) remark that to many biologists "Transformation seems too much like "data grinding." When you learn that a statistical test may be made significant after transformation of a set of data, though it would not have been so without such a transformation, you may feel even more suspicious. What is the justification for transforming the data?"

The justification for transforming data, the decision about whether to transform in a particular instance, and the choice of transformation are best approached through the following sequence, which is illustrated later by application to our effluent impact example.

1. Are there serious violations of assumptions? The most serious violation which can be corrected *after* the data have been collected is heterogeneity of error variances (Section 2.1.7). Cochran (1947) indicates that loss of efficiency in estimation of treatment effects and loss of sensitivity (higher Type II error rate β) in significance tests are the consequence of this violation. Overall F-tests are probably affected the least. However, with unequal sample number the Type I error rate α may also be affected (Glass et al 1972). If the groups with the larger variances tend to have the larger sample numbers, the test is more conservative. For example, you may think you are testing at $\alpha = 0.05$ while actually testing at $\alpha = 0.02$. If the groups with the larger variances tend to have the smaller sample numbers, the test is more liberal, so that a test nominally at $\alpha = 0.05$ may actually be at $\alpha = 0.10$. Obviously this is the case to worry about. As previously mentioned (Section 2.3.8), decisions about sampling design and methodology (e.g., stratification, choice of sample unit size) based on results of preliminary sampling should have minimized any violation of this assumption. If the remaining heterogeneity of error variance can be cured, less serious violations that tend to be correlated with it will probably also be cured or at least reduced.

2. Is the assumption of homogeneous error variances a tenable one for the data in question? Must we reject the H_0: "homogeneity of variance"? There are a number of possible approaches at this point. Often formal tests are not necessary. Skillful examination of scatter plots and histograms (Section 2.1.7) of the raw data (or residuals) and of the sample variances and covariances for each group usually provides an adequate basis for deciding whether variance heterogeneity is a problem (see Section 3.9). Bartlett's test, which is described and illustrated by an example in Steel and Torrie (1960), is probably the most commonly used formal test of the null hypothesis of homogeneity of variance. When it is used, a significant X^2-statistic should *not* be taken to mean that there would necessarily be serious consequences on tests of significance about group mean differences

if the raw (untransformed) data were used for these tests. It should rather be interpreted as an indication that some improvement may be possible, that there is some heterogeneity of error variance, and that it is worth investigating whether that heterogeneity can be reduced by transforming the data. As Harris (1975) comments, Bartlett's test "has high power to detect departures from homogeneity which are too small to have any appreciable effect on the overall F-test." For two groups the ratio of the larger to the smaller sample variance may be treated as an F-statistic to test against H_0: "equal within-group variances," as described in Sokal and Rohlf (1973). Also described there is the "F_{max}-test" for more than two groups, where F_{max} is the ratio of the largest to the smallest within-group variance. A table of critical values for F_{max} is provided. This test can be used as an alternative to Bartlett's test.

3. If H_0: "homogeneity of variance" is rejected, will a transformation— *any* transformation—reduce the heterogeneity? We must ask, what can a transformation do? It can change the scale of measurement so that the variances are independent of the mean values. By removing functional dependence of the variance on the mean (which is common in data obtained by sampling distributions of organisms) a transformation removes any heterogeneity of variances that is related to differences in means among groups. Differing within-group variances that result from other, perhaps unknown, causes cannot be eliminated by a transformation. Steel and Torrie (1960) refer to these two types of error heterogeneity as "regular" and "irregular," respectively. The irregular type may result from outliers, which are gross errors or extremely atypical samples, or from some spatial heterogeneity that should have been handled by a better sampling design (e.g., stratification). Some methods for detecting outliers are described in Section 3.9. For testing against the $H_0 : \sigma^2 \neq f(\mu)$, that the variance is independent of the mean, a number of approaches are possible. A simple procedure for testing this hypothesis, and also for choosing the correct transformation if the hypothesis is rejected, is based on Taylor's power law (Taylor 1961), which is well described and illustrated by Southwood (1966). This law states that for most field distributions of organisms the variance-mean relationship can be described by the model $\sigma^2 = a\mu^b$, which can be expressed in linear form as $\log \sigma^2 = \log a + b \log \mu$. Using the sample statistics s_j^2 and $\overline{X}_j$ as the estimates of σ_j^2 and μ_j respectively for each group j, the constants a and b may be estimated by linear regression. The test for significance of the regression (the test against $H_0 : b = 0$) is also the test against $H_0 : \sigma^2 \neq f(\mu)$. If $b \neq 0$, the estimated nonzero b value indicates the appropriate transformation as $Z = X^{1-(1/2)b}$, unless $b = 2$ in which case $Z = \log X$. With field population data the slope b will

often be in the range 1 to 2 (see Taylor 1961) where $b = 1$ would indicate a square root transformation and $b = 2$ a logarithmic transformation. In most cases it makes sense to choose one or the other rather than something in between, although Reys (1971) and Williams and Stephenson (1973) find a cube-root transformation ($Z = X^{1/3}$, which implies a $\sigma^2 = a\mu^{4/3}$ relationship), to be useful. See below for related discussions. Papers by Healy and Taylor (1962) and Box and Cox (1964) also deal with the estimation and use of power law transformations.

4. Transform the data according to whichever model is chosen ($Z = \log X$, say) and proceed with the statistical analysis. All means, standard deviations, confidence limits, and tests of significance should be calculated for Z. If means and confidence limits in the original units are desired, as the *last* step back-transform the mean, the upper confidence bound, and the lower confidence bound from Z to X units. Before using the transformed Z data in the statistical analysis you could redo Bartlett's test to test for any remaining heterogeneity of variance (presumably of the irregular type) in the Z data. However, there seems little point to it, since Bartlett's test is not an appropriate indicator of whether the heterogeneity is large enough to require more drastic measures such as nonparametric methods.

Each of the four remedies is now considered in more detail.

The word "transformation" should be interpreted broadly to include more than, say, taking the logarithm or the square root of each value. For example, the data may be transformed to ranks and then by the Fisher and Yates (1967) transform to average deviates of the rth largest of samples of n observations drawn from a normal distribution with unit variance. The latter step also makes the rank values independent of the number of samples. Such transformed rank values perform adequately in parametric statistical analyses such as ANOVA if there are not too many tied ranks (Marriott 1974).

It is often useful to transform correlated multivariate data by finding the linear additive functions of the original correlated variables which allocate the greatest possible variation to the fewest possible new uncorrelated variables. This is called principal components analysis (PCA) and is equivalent to a rigid rotation of the axes about the origin so that they "line up" as much as possible with the trends in the data (Cooley and Lohnes 1962, 1971; Harris 1975; Marriott 1974). In such analyses as multiple regression and discriminant analysis, tests of the significance of the original predictor variables assume the independence of those variables (probably the most ignored assumption in statistical analysis). PCA can be used to transform such correlated variables to a new set of independent variables that retain all of the information in the old set for subsequent statistical analysis. See Section 3.4 for further discussion and an example.

Simple data coding may be thought of as a transformation, for example $Z_i = (X_i - \overline{X})/s_i$, which standardizes the data so that observations on each variable i have zero mean and unit standard deviation. Some workers do this as the first step in virtually every statistical analysis procedure. Such standardization, however, has implications for interpretation of the analysis results, which should also be considered. Where the relative magnitude of variables is of primary interest—for example, when using species abundance data to estimate percentage composition—standardization is often appropriate (Austin and Greig-Smith 1968). The results of an analysis may depend on the scale of the original measurements unless the data are standardized to ensure scale-independence (Jolicoeur 1963a, Gower 1967). Some of the implications of data standardization are discussed by Orloci (1967) and Jolicoeur and Mosimann (1960). Noy-Meir et al (1975) provide a good review paper on the subject. Logarithmic transformation of data *by itself* has the effect of making the observations scale-independent (Marriott 1974, Jolicoeur 1963a,b), which is usually the best approach. Standardization of log-transformed data accomplishes nothing of value and may complicate the interpretation of the results (Jolicoeur 1963b).

Coding the data so that most values lie between 1 and 10, or at least below 10, minimizes rounding error (Section 2.1.9). Coding by choice of units of measure can influence the effect of subsequent transformations. For example, Rounsefell and Dragovich (1966) code plankton counts as numbers per centiliter, since in using milliliters the transformation $Z = \log (X + 1)$ overcorrected for very low counts (see below for discussion on the use of this transformation). Finally, coding the data in the range 1 to 10 minimizes any violation of the additivity assumption. Data obtained by sampling organism distributions are often characterized by multiplicative rather than additive effects. Since $2 \times 2 = 2 + 2$, it is apparent that in this range the results obtained by multiplicative and additive models would differ the least.

The choice of a specific transformation is, as mentioned above, based on the variance-mean relationship, which in turn is a function of the spatial distribution. Even when the distribution being sampled is strictly random (the Poisson distribution), dependence of the variance on the mean is a problem ($\sigma^2 = \mu$) and a transformation ($Z = X^{1/2} = \sqrt{X}$) is appropriate. Many field distributions of organisms are highly aggregated ($\sigma^2 >> \mu$) and approach the logarithmic series distribution where $\sigma^2 \alpha \mu^2$ and $\sigma \alpha \mu$. Here the logarithmic transformation is appropriate. Thus arises the tendency for b in Taylor's power law $\sigma^2 = a\mu^b$ to lie in the range 1 to 2 for field data. There is a vast literature on spatial distributions, and a source of confusion is the fact that there are two entirely different motives for modeling spatial distributions of organisms. Our interest is in fitting a model that will aid in the decision of whether transformation is necessary and which transformation is appropriate. For this we require a general model (Taylor's power law, say) that fits most commonly encountered organism distributions. Many spatial distribution models are, however, designed to explain the causes of particular spatial distributions, for example, to test against H_0: "an adequate explanation of the observed distribution is random dispersal of propagules each of which gives rise to colonies whose individual populations are of random sizes" (the Thomas distribution). Such models *must* be very specific (hence there are very many of them) or the power of the test against the particular H_0 is low. Models used to aid in transformation, such as Taylor's power law and the negative binomial distribution, fit most anything. Southwood (1966) lists five different and biologically realistic ways in which the negative binomial may be generated, for example. Obviously the result of a test against H_0: "negative binomial" would be useless for inferring the biological process that brought it about. As with Taylor's power law, however, the estimated parameters of the distribution provide a basis for transforming the data so that they are statistically well-behaved.

The negative binomial (NB) distribution is a generalization of the Poisson and the logarithmic series distribution, as can be seen from the variance-mean relationship determined by it: $\sigma^2 = \mu + \mu^2/k$. As k becomes very large, the Poisson distribution is approached; as k becomes very small, the contribution by the μ term becomes insignificant and the logarithmic series distribution is approached. Therefore the range $k = \infty \to 0$ in the NB distribution corresponds to the range $b = 1 \to 2$ for Taylor's power law. Anscombe (1948, 1950) shows that there is an exact transformation based on the NB, the inverse hyperbolic sine square root transformation (see the ordinate in Figure 2.7), and he compares the properties of the NB with other distributions. Bliss (1953) discusses procedures for fitting the NB distribution in addition to those mentioned by Anscombe. Taylor (1953) applies the NB to fisheries trawl catch data and shows that smaller sample unit size is more efficient if the NB distribution is the model (Section 2.3.8). Chatfield (1969) presents some rapid estimation procedures, and Southwood (1966), Elliott (1977), and Poole (1974) provide thorough treatments of the fitting and use of the NB distribution. There seems little to choose between the NB and Taylor's power law, as both appear to provide an adequate fit to a wide range of data from field distributions of organisms. Kuno (1972) argues that Taylor's power law

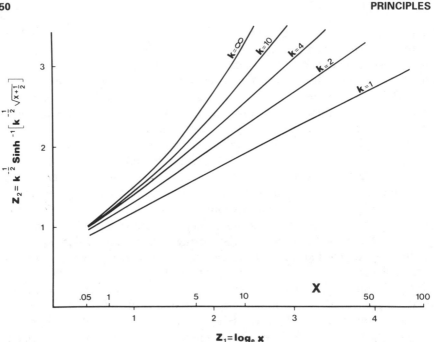

FIGURE 2.7   The relationship between the exact negative binomial transformation Z_2 with different values of k and the logarithmic transformation Z_1.

is biologically unrealistic under certain conditions, but his objections are probably of more theoretical than practical interest. The NB model is particularly useful in sequential sampling designs, which are discussed below.

In practice most data from samples of field distributions of organisms may be transformed as $Z = \log (X + 1)$, which allows zero values to be used whereas $\log X$ does not. Also $\log (X + 1)$ is a better transformation for small values. Steel and Torrie (1960) comment that "it is desirable to have a transformation which acts like the square root for small values and like the logarithmic for large values," and they conclude that the $Z = \log (X + 1)$ transformation does just that. For most field data the value of k in the NB is 2 or less and the value of b in Taylor's power law is close to 2, suggesting that the logarithmic transformation is generally applicable. Cassie (1962) concludes that "In practice, there seems to be little loss of precision. . . . Since the convenience of the log transform outweighs most other considerations, it is doubtful whether many ecologists would find time to employ any other." For discussion of some implications of this apparent general applicability of the logarithmic series distribution for estimation of necessary sample number, see Section 3.8. In the same vein

Bartlett (1947) discusses the exact NB transformation, concluding that $Z = \log (X + 1)$ is likely to be good enough in many cases because it is linear with the NB transformation "for values of λ [where $\lambda = k^{-1}$] which appear likely in practice." Many environmental biologists have gone through the cycle, as I did, of discovering the NB distribution, calculating exact NB transformations for masses of field data, and finally finding the near-linearity with the much simpler logarithmic transformation that Bartlett describes. This is shown in Figure 2.7. Another *a priori* argument for logarithmic transformation of species abundance data is that temporal change in species abundances at constant percentage rates is converted to linear form and therefore is amenable to regression or other linear model statistical analysis. See Figures 2.8 and 2.9 as examples (taken from Green and Hobson 1970). Figure 2.10a,b (taken from Cassie and Michael 1968) shows the normalizing effect of a logarithmic transformation on a bivariate species abundance plot. Four examples of the use of the $Z =$

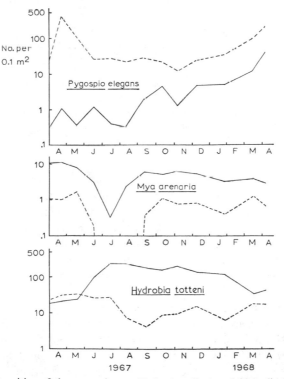

FIGURE 2.8 Densities of three species at 60 (broken line) and 90 (solid line) cm above mean low water in the Barnstable Harbor, Massachusetts, intertidal community between March 1967 and April 1968. The ordinate is on a logarithmic scale. Reproduced with permission from Figure 3 of Green and Hobson (1970).

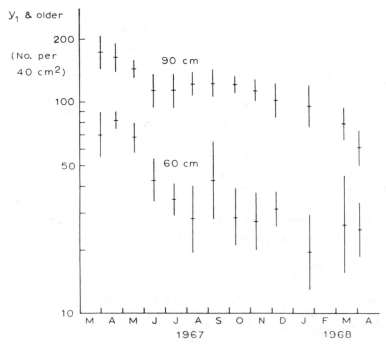

FIGURE 2.9 Changes of densities of the 1966 (Y_1) and older year classes of the clam *Gemma gemma* at the 60 and 90 cm tidal levels. See legend for Figure 2.8. Horizontal bars represent means, and vertical bars represent .95 confidence limits for the estimates of the means. The ordinate is on a logarithmic scale. Reproduced with permission from Figure 5 of Green and Hobson (1970).

log (X + 1) transformation are in papers by Rounsefell and Dragovich (1966), Buzas (1967), Hughes (1971), Bodiou and Chardy (1973), and Cassie and Michael (1968). The last of these, and Green (1971a), provide examples of logarithmic and other transformations of environmental as well as biological variables. Concentration variables that vary over orders of magnitude are often logarithmically transformed. The variable pH is nothing but a logarithmic transformation of the hydronium ion concentration, for example.

Other transformations are applicable in environmental studies, and some general references are Elliott (1977), Steel and Torrie (1960), Cassie (1962), Bartlett (1947), and Barnes (1952). Transformations for multivariate data are considered by Andrews et al (1971) and Cassie and Michael (1968). I will mention only one other transformation here—the $Z = \sin^{-1} \sqrt{X}$ transformation, which is often applied to proportions. It is probably overused. Steel and Torrie (1960) point out that proportions do not require transformation unless they are outside the range 0.3 to 0.7. Practicing

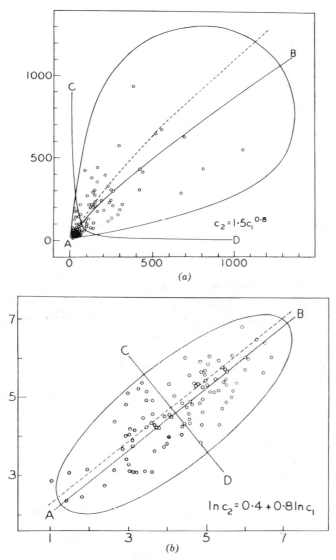

FIGURE 2.10 (a) A scatter plot of simulated abundances of two species (c_1 and c_2). Lines A–B and C–D correspond to the lines labeled similarly in (b), where the axes c_1 and c_2 are logarithmically transformed. The closed curve is the .99 probability envelope, which is an ellipse in (b). Reproduced with permission from Figure 4 of Cassie and Michael (1968).

ecologists have often found that this arcsin square root transformation has little effect (Loucks 1962, Dale 1964, Orloci 1966). It probably does no harm either.

Nonparametric methods are more likely to be useful when the entire study is planned with their use in mind, rather than as some kind of salvage operation for unexpected intractable data. Presence-absence or rank data are often appropriate for ecological studies, both because more (not less) information per unit time and cost expended may be collected in this form and because statistical models based on changes in frequencies or in rank order may be more appropriate (see Section 3.4). Levins (1966) argues that precision is the attribute of the perfect model which should be sacrificed, rather than generality or realism (see Section 2.1.4). Hypotheses about which species will increase and which decrease, for example, can lend themselves to adequate and robust tests of important hypotheses. Often it is best to leave out of the hypotheses any prediction about the magnitudes of the increases or decreases. Problems with biased estimation of populations and communities from samples were discussed in Sections 2.1.5 and 2.3.6. The biases of particular sampling methods are often impossible to eliminate, and it has been observed many times that the number of species, species identities, and their rank abundances are much less affected by sampling bias and other sources of noise than are quantitative species abundances. See Section 3.4 for further discussion on decisions about the form in which the data should be collected or coded after collection. Probably the most useful area for statistical methods based on presences and absences or on ranks is that of analysis and interpretation of multivariate (including bivariate) relationships. The linear model assumption on which most standard methods (e.g., principal components analysis, factor analysis, canonical correlation analysis) are based is often inappropriate for ecological data. Relationships between species abundances and environmental factors, for example, are rarely linear and often not even monotonic. In recent years there has been active development of rank or order methods analogous to most of the multivariate linear model analyses, including factor analysis and other ordination procedures, discriminant analysis, and canonical correlation analysis. See Section 3.4.2 for further discussion about alternatives to linear models for nonlinear ecological data.

In addition to the standard references given previously, a manual by Wilcoxon and Wilcox (1964) illustrates some rapid approximate statistical tests. Some simple tests for location and trend are described by Cox and Stuart (1955).

Sequential sampling, though not often appropriate, can be extremely useful. Developed largely during World War II for quality control in producing war materials, it has since found application in many areas including routine survey work in environmental studies. The principal advantage of sequential sampling is that the statistics are done once, based on preliminary sampling, and in advance of the study itself. In the actual study, sampling continues until a decision is made so that prior estimation of adequate sample number is unnecessary. Decisions related to specified tests of hypotheses with known error risks are converted to graphical format, which can then be used over and over as long as the assumptions that went into the original sequential sampling design remain satisfied. The last qualifier is the catch. On the one hand, sequential sampling is very flexible because designs can be based on virtually any sampling distribution, such as the normal, Poisson, binomial, or negative binomial. However, a given design is useful only so long as the sampling distribution on which it was based does not change, and spatial distributions of organisms do change throughout the course of a study. If the sequential sampling graphs had to be redesigned frequently, the primary advantage of using this approach would be lost. The negative binomial distribution has been successfully applied to sequential sampling, for example by Oakland (1950) for sampling whitefish for parasites and by Morris (1954) for spruce budworm egg surveys. The difficulty is that not only the NB distribution, but also the *particular* NB distribution, must hold—that is, the degree of nonrandomness as measured by the parameter k must not change. Also, sequential sampling tests are generally less efficient than the standard ANOVA-type tests, there is little advantage to sequential sampling unless the cost per sample collected is much the same regardless of how many are collected at the same time, and the time required for sample processing cannot be large. If most of the cost is in the trip to the location where the samples must be collected rather than in the cost-per-sample after arriving on site, or if much time must pass between sample collection and having the results, then sequential sampling is not the way to go.

Southwood (1966) and Poole (1974) discuss sequential sampling and give examples. An elegant example that is useful for binomial data is the closed sequential test design of Cole (1962) which is shown in Figure 2.11. It is a sequential sampling version of the sign test (Steel and Torrie 1960). Beginning in the box marked X in the lower left-hand corner, each sample collected in turn represents a move up or to the right depending on whether $A > B$ or $B > A$ for that sample. Sampling continues until one of the three regions is entered and one of the three corresponding decisions is made. The null hypothesis is H_0: $A = B$, and the design is such that $\alpha = 0.05$ with this Type I error risk split evenly between $A > B$ and $B > A$. The level of β varies with true magnitude of departure from $B = A$, and is

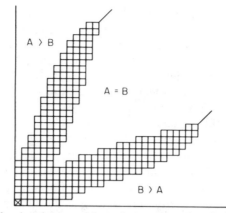

FIGURE 2.11 The closed sequential test design. Reproduced with permission from Figure 1 of Cole (1962).

tabulated by Cole. Each sample might be a trial run of two randomly selected individuals of the same species against effluent water (A) on the one hand and control water (B) on the other hand. A move up or to the right would depend on the relative levels of some criterion variable in the two cases. Time to death or disability, oxygen consumption rate, or activity are possible examples. The H_0 would be that the effluent water did not affect the organism differently than did the control water. Another possible use would be for selection of an indicator species by subjecting a randomly selected individual from each of two species, A and B, to the same toxicant and seeing which died first, then selecting another pair, and so on. The H_0 in this case would be that the two species do not differ in their sensitivity to the toxicant. The possibilities are endless.

Another way in which sequential designs can be used is to combine several tests so that mean levels can be assigned to arbitrary categories. For example, one can construct a sequential sampling graph for deciding whether a mean of samples from a negative binomial distribution with $k = 2$ is greater than 1 (the H_A) by 100 percent, say, or less than 1 (the H_0), with error risks $\alpha = \beta = 0.05$. If the decision lines for H_A's and H_0's, respectively, of 20 and 10, and 200 and 100 are then calculated for the same distribution and error risks, they can be combined as shown in Figure 2.12 to provide a quick "sorting by order of magnitude" sequential sampling procedure. If it were known that the degree of aggregation (i.e., the level of k) changed in some predictable way with mean density then each of the pairs of lines could be constructed for different k values.

A sequential design for sampling until some specified precision for the estimate of the mean value is achieved is described in Section 3.8.

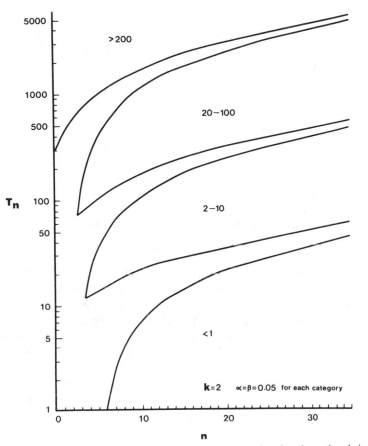

FIGURE 2.12 A sequential sampling design for placing species abundance levels into order of magnitude categories given a negative binomial distribution with $k = 2$ and error levels $\alpha = \beta = 0.05$ for each category. T_n is the cumulative total number of organisms sampled in n samples.

The use of simulation for testing hypotheses with data that violate the assumptions of standard methods was discussed in Section 2.1.7. Testing in this way is illustrated below with our effluent impact example. The SAS statistical package was mentioned, but any computer package or programmable calculator with functions for random number generation will allow data simulation for this purpose. Capra and Elster (1971) describe a method for simulating multivariate normal data with specified means, standard deviations, and correlations. This method can be used to generate small multivariate data sets on programmable calculators (see Section 4.1).

Two good examples of determination of spatial distributions for choosing appropriate transformations, and subsequent annalysis on that basis, are given by Bishop and De Garis (1976) and Reys (1971). Bishop and De Garis sample species distributions, transform random and aggregated distributions, and then use the transformed data to calculate proper confidence limits. Reys analyzes the distributions of a large number of benthic species near Marseilles, noting that "The display of the type of distribution allows a better estimation of the density and shows what transformation must be applied to the original data before application of classical statistical methods." He finds that 88 percent of the species were significantly aggregated with variance dependent on the mean. Clump, or patch, size seemed to depend on the feeding habit of the species. Fitting of the negative binomial (with estimation of a common k), power law estimation of transformation, and graphical displays of the distributions are extensively presented.

Choice of the statistical analysis model was discussed in Section 2.1.6. The DECISIONS section, which follows, describes the basis for allocation of an environmental study to one of several main sequences, and in the SEQUENCES section appropriate statistical models for each are considered and illustrated. For our effluent impact example we use a 2×2 areas-by-times factorial analysis of variance design, where the test against H_0: "interaction" is the test against the H_0: "no change in the density of the criterion species in the impacted area that does not also occur in the control area" (see Sections 2.3.1, 2.3.4, and 2.3.8). Discussion of this as an optimal impact study design is in Section 3.2. Calculations for the factorial ANOVA are presented with examples in Steel and Torrie (1960) and Sokal and Rohlf (1969).

It was previously (Section 2.3.8) decided that $r = 6$ random replicate samples would be collected in each area-by-time combination for a total of 24 samples. The data are given in Table 2.1 as the original quantitative observations, as ranks, and in binary form (as above or below the median).

First we test against the H_0: "sample variances for each of the four groups (the areas-by-times combinations) are estimates of a common within-group variance." Sample variances and means for the four groups are:

	Before		After	
	Control	Impact	Control	Impact
s^2	255	792	117	25.5
$\overline{X}$	45.8	52.5	41.0	10.5

Table 2.1 Species abundances, rank abundances, and dichotomized abundance for criterion variable species Y in control and impact areas at before-impact and after-impact times

Time	Control area						Impact area					
Before-Impact												
b Quantitative	36	67	30	65	40	37	24	60	24	41	95	71
Ranks	12.5	3	16	4	9	11	17.5	5	17.5	8	1	2
Binary	1	1	0	1	1	1	0	1	0	1	1	1
After-impact												
a Quantitative	36	32	49	59	38	32	8	8	20	12	9	6
Ranks	12.5	14.5	7	6	10	14.5	22.5	22.5	19	20	21	24
Binary	0	0	1	1	1	1	0	0	0	0	0	0

Bartlett's test yields:

$$X^2 \ (3 \ \text{df}) = 4 \ (r-1) \ln \bar{s}^2 - (r-1) \sum_1^4 \ln s^2$$

$$= 20 \ln 297 - 5(5.5413 + 6.6746 + 4.7622 + 3.2387)$$

$$= 113.87 - 101.08$$

$$= 12.79 \ (p < 0.01).$$

If the X^2-statistic were marginally significant, a correction factor should be applied, which would lower it slightly. We reject H_0 and conclude that the within-group variances do differ. The same decision would have resulted from the F_{max} test, since the ratio of the largest to the smallest variance is 31.1 and the .95 critical value of F_{max} is 13.7 [see Sokal and Rohlf (1973), Table VII].

It should be emphasized that where I present more than one test of the same H_0, as above for H_0: "homogeneity of variances," it is for purposes of illustration only. I am *not* implying that selection of the desired answer from several tests of the same H_0 is legitimate (see Sections 2.1.7 and 2.3.10).

To what extent is the heterogeneity of variances dependent on the mean? Given this type of data (species abundances) we would expect *a priori* that a $Z = \log (X + 1)$ transformation would be appropriate. Therefore we proceed on that basis unless there is evidence that it is inappropriate to do so. Since there are only four groups, Taylor's power law must

Table 2.2 The factorial analysis of variance table for abundance of criterion variable species Y

Source	df	SS	MS	F
Between times	$t - 1 = 1$	3.69	3.69	
Between areas	$a - 1 = 1$	2.47	2.47	
Interaction	$(t - 1)(a - 1) = 1$	2.90	2.90	18.5**
Error	$at(r - 1) = 20$	3.13	0.157	
Total	$atr - 1 = 23$	12.19		

**$p < 0.01$.

be fitted as a linear regression $\ln s^2 = a + b \ln \overline{X}$ to only four points. Estimates of a and b respectively are -0.991 and 1.75, yielding $s^2 = 0.37$ $\overline{X}^{1.75}$ for Taylor's power law. The exponent is close to 2, which agrees with the logarithmic transformation. With only 2 error degrees of freedom the regression is not significant (H_0: $b = 0$ cannot be rejected) despite a coefficient of determination of 0.82, which implies that dependence of the variance on the mean explains 82 percent of the heterogeneity of variance.

Using $Z = \ln(X + 1)$ we calculate the factorial ANOVA, and the results are shown in Table 2.2. With this transformation the data are in effect standardized, and the values lie within the range 1 to 10 (see discussion re coding earlier in this section). The F-test for interaction is based on $F(1, 20\ df)$, and its significance (at $p < 0.01$) invalidates tests for main effects (against H_0: "abundances similar in the two areas at both times" or H_0: "abundances similar at the two times in both areas"). See Section 4.1 for discussion of tests of these hypotheses. Here we do conclude a significant interaction, and consequently significant impact effects. This agrees with the known properties of the data (Section 2.2.2). The .95 confidence limits in the transformed Z units for the impact area after impact (AI) are

$$\overline{Z}_{AI} \pm t_{.975}(20\ df) \sqrt{\frac{\text{error mean square}}{6}}$$
$$= 2.3754 \pm (2.086)(0.1618)$$
$$= 2.3754 \pm 0.3374$$
$$= 2.0380 \text{ to } 2.7128.$$

In the original X units (number per m²) the mean and .95 confidence limits for the impact area after impact are

$$\text{mean} = e^{2.3754} - 1 = 9.75,$$

$$.95 \text{ confidence limits} = e^{2.0380} - 1 \text{ to } e^{2.7128} - 1$$

$$= 6.68 \text{ to } 14.1.$$

Means and confidence limits would be calculated similarly for the other three area-by-time combinations. Note that the .95 confidence limits would be nearly symmetric about the mean in X units if plotted on a log scale, as they are for the data shown in Figure 2.9. (They would be exactly symmetric for a $Z = \log X$ transformation.)

Now let us look at alternative analyses we might have performed to test the same hypothesis (H_0: "no areas-by-times interaction"). Nonparametric analogues to simple one-way ANOVA do exist, but for the factorial design the best approach is to replace the observations by ranks and then apply the Fisher and Yates (1967) transform to convert them to average normal deviates, which we then use in the same factorial ANOVA as illustrated above. The results are shown in Table 2.3. The F-test for interaction is again significant at $p < 0.01$, although with a lower F-value. The replication of $r = 6$ samples per area-by-time combination, which was estimated for quantitative data, was also more than adequate for an analysis based on ranks. However, the estimation of necessary replication in practice should be based on the analysis and the type of data to be used.

The most efficient conversion into binary data is into equal halves, and in this case we have used above or below the median. Median values were decided by the flip of a coin. In practice binary data are usually presence-absence, and the efficiency can be maximized by choosing a sample unit size that results in roughly half presences and half absences. Species that are mostly present or mostly absent contribute little information. Again, the importance of planning the sampling, the coding of the data, and the

Table 2.3 The factorial analysis of variance table for transformed rank abundances of criterion variable species Y

Source	df	SS	MS	F
Between times	$t - 1 = 1$	6.44	6.44	
Between areas	$a - 1 = 1$	2.12	2.12	
Interaction	$(t - 1)(a - 1) = 1$	4.07	4.07	8.86**
Error	$at(r - 1) = 20$	9.20	0.460	
Total	$atr - 1 = 23$	21.83		

**$p < 0.01$.

statistical analysis in advance and as one integrated sequence should be strongly emphasized (Section 2.3.8).

The data in binary form are:

Species	Before		After	
	Control	Impact	Control	Impact
1	5^a	4^b	4^c	0^d
0	1^e	2^f	2^g	6^h

Even binary data of this type could be used (as 0/1 values) in an analysis such as the factorial ANOVA. Surprisingly good results can be obtained if sample number is large. Glass et al (1972) conclude that normal distribution statistics are robust even with dichotomous data if groups have equal sample number and there are enough error degrees of freedom.

However, statistical analyses designed for contingency table data are generally most appropriate. The calculations that follow are after Steel and Torrie (1960, p. 384). This problem may also be approached by way of log linear and linear logistic models (Cox 1970, Fienberg 1970). See also section Section 3.4.1. Fienberg emphasizes application in ecology and discusses analyses of contingency tables as an alternative to analyses by ANOVA with transformations. Obviously this particular analysis would not be of practical value with so few samples, and the X^2 approximation would not be valid. However, we proceed for the sake of the example.

Here we wish to test a three-factor (or second order) interaction. If the species occurrence, as 0/1, responds to an impact-related change that occurs between times, the null hypothesis is H_0: "any time-by-species interaction differs between the areas by no more than random variation." Let $f_{i=1,8}$ be the true frequencies for cells 1 to 8 where $a, b, ..., h$ are the observed values. If H_0 is true, then

$$\frac{f_1 f_7}{f_3 f_5} = \frac{f_2 f_8}{f_4 f_6} \quad \text{and} \quad f_1 f_4 f_6 f_7 = f_2 f_3 f_5 f_8.$$

Expected frequencies are given by

$$(a + x)(d + x)(f + x)(g + x) = (b - x)(c - x)(e - x)(h - x)$$

where x is the deviation of the observed from the expected value in this 1 degree of freedom situation. For our data

$$(5 + x)(0 + x)(2 + x)(2 + x) = (4 - x)(4 - x)(1 - x)(6 - x)$$

which gives the cubic equation

$$x^3 - 2.25x^2 + 7.5x - 4 = 0.$$

Since $x = 0.6160$,

$$X^2 \ (1 \ df) = x^2 \ \Sigma \ (\text{expected})^{-1}$$

$$= (0.6160)^2 \left(\frac{1}{5.616} + \frac{1}{0.616} + \frac{1}{2.616} + \frac{1}{2.616} \right.$$

$$\left. + \frac{1}{3.384} + \frac{1}{3.384} + \frac{1}{0.384} + \frac{1}{5.384} \right)$$

$$= 2.26.$$

Of course, this is nonsignificant. The correction for continuity, which should be applied because there is only 1 degree of freedom, would reduce the X^2 value even further.

The null hypothesis tested by the factorial ANOVA (H_0: "no areas-by-times interaction") is based on a model that assumes homogeneity of variances. After we rejected the homogeneity of variances assumption we proceeded to stabilize the variance with a transformation. However, we could instead compare the interaction F-value obtained from the factorial ANOVA using the untransformed data, with F-values that would be obtained using "H_0 data" without homogeneous variances. That is, we can simulate a large number of data sets that have the mean value for all groups ($\mu - 37.45$) but variances as estimated for each group. A normal within-group distribution is assumed, for lack of any evidence to the contrary. (Of course, we *know* that the data are lognormally distributed (Section 2.2.2), but in practice the normality assumption could not be rejected with such a small data set. It would not be a serious violation in any case.) The F-value for interaction calculated for the raw data is $F_{int} = 6.97$. Of 100 simulated H_0 data sets with $\mu = 37.45$ and variances $\sigma_{BC}^2 = 255$, $\sigma_{BI}^2 = 792$, $\sigma_{AC}^2 = 117$, and $\sigma_{AI}^2 - 25.5$ (where BC – before control, BI = before impact, AC = after control, and AI = after impact) not one yielded an F_{int} as high as 6.97. The .95 confidence limits on a proportion estimated from zero occurrences out of 100 observations do not include 0.05. We therefore conclude that H_0: "no areas-by-times interaction" should be rejected.

2.3.10   Having chosen the best statistical methods to test your hypothesis, stick with the result. An unexpected or undesired result is not a valid reason for rejecting the method and hunting for a "better" one.

Pin your hopes on the best statistical method you can find for testing the hypothesis in question using the best data you can obtain. Accept that either H_0 or H_A are possible outcomes. Anyone who says, "That difference is almost significant!—I'll bet this other method will show it," is not testing falsifiable hypotheses (see Section 2.1.2). He or she is engaged in a ritual of self-deception, and is like one who roots about in masses of data looking for the 5 percent of correlations that are significant at $p < 0.05$ (see Section 2.1.6). As in the fable, the statistical method is only the messenger, which may bring bad or good news. Obliterating the messenger does not change the reality. It has always been possible to arrange things so that you will be told what you want to hear.

Three

DECISIONS

3.1 INTRODUCTION

In an environmental study the sampling design and the statistical analyses that flow from it can be set in a spatial-by-temporal framework, such as that shown in Figure 3.1. Options within this spatial-by-temporal framework are shown in Figure 3.2. Samples can be collected at one or more times and at one or more locations. They can be taken before an environmental impact, after it, or both. In the spatial dimension, sampling can be done in one area that is assumed to be homogeneous with respect to degree of impact, or there may be several locations within each of two or more areas chosen to represent different degrees of impact. Within each unique combination of location and time one or more samples should be taken to determine levels of one or more variables. Ideally (see Sections 2.3.1 and 2.3.2), an equal number of randomly allocated replicate samples should be taken at each location and time. It is often appropriate to partition the variables into two sets. One set contains biological variables such as species abundances, and the other set environmental variables that may be related in some explanatory manner to the biological set. These sets correspond to predictor and criterion variable sets, but whether the biological variables are the predictor or the criterion set depends on the purpose of the study. If one wishes to explain change in species composition by changes in environmental variables, the biological variables represent the criterion set and the environmental variables the predictor set. If the abundances of certain indicator species are used as predictors of, say, a particular type and intensity of chemical pollution, the situation is reversed.

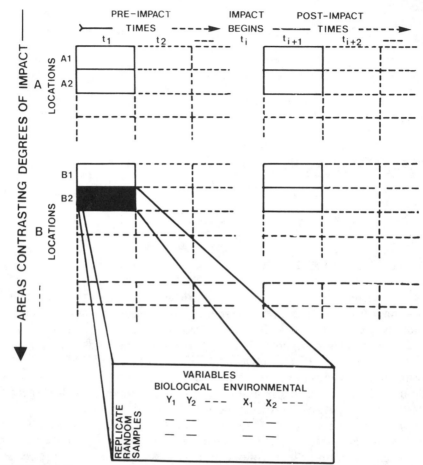

FIGURE 3.1   The spatial-by-temporal framework for sampling design and statistical analysis in environmental studies.

The total data set for all locations and times can be equated to the data matrices and their corresponding statistical models in Figure 3.3. Partition of variables into two sets is as described above, and partitions of samples are in either or both of the two dimensions represented by levels of the controlled variables location and time. Appropriate statistical analysis often takes the form of a sequence of different analyses applied to different partitions of the data matrix. For example, the first analysis step for a data matrix partitioned into sets of variables (as in the second row of Figure 3.3) might be reduction of the criterion (= biological in this case) variable set to fewer variables using a procedure such as principal com-

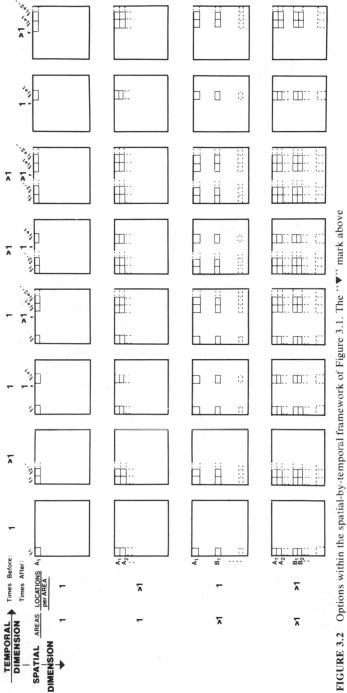

FIGURE 3.2 Options within the spatial-by-temporal framework of Figure 3.1. The "▼" mark above columns 3 to 8 represents the beginning of impact.

ponents analysis (Section 3.4.3). For this step in the analysis the data submatrix is as in the first row of Figure 3.3. The second step might then be some statistical procedure that relates the reduced criterion biological variable set to the predictor environmental variable set, an analysis corresponding to the second row of Figure 3.3.

Applied environmental studies can be classified into baseline, monitoring, or impact studies. These terms are often used loosely. The following definitions apply to their use in this book. A *baseline* study is one in which data are collected and analyzed for the purpose of defining the present state of the biological community, the environment, or both. Usually some environmental change is anticipated, though both the nature of the change and the time of its occurrence may be unknown. The present state may contain patterns of spatial and temporal variation, which would have to be incorporated into the baseline model. An *impact* study is one whose purpose is to determine whether a specified impact causes change in a biological community and, if it does, to describe the nature of that change. One may be able to obtain both before- and after-impact data or only after-impact data. There may or may not be a control area. The nature of the impact and the fact of its occurrence is always known. A *monitoring* study has the purpose of detecting a change from the present state. If the data used are from sampling to detect change in the biological community, it is a *biological monitoring* study. Baseline data must be available to provide a standard against which to detect a change. The nature of the change may be defined very specifically in terms of a particular type and degree of impact or it may not be defined at all. For monitoring studies to be most sensitive in detecting change, the change should be clearly defined on the basis of results from previous impact studies. This provides a specific H_A against which to test the H_0 of no change.

3.2 OPTIMAL IMPACT STUDY DESIGN

An impact study is best designed when it judges impact effects against previously collected baseline data, and it is best used when the results provide the basis for subsequent monitoring to detect future impacts of the same type. An optimal impact study design therefore provides the best starting point for discussion, and all other designs may be treated as suboptimal designs that have one or more missing parts.

There are four prerequisites for an optimal impact study design. First, that impact must not have yet occurred, so that before-impact baseline data provide a temporal control to which the after-impact data can be contrasted. Second, the type of impact and time and place of occurrence

DATA MATRICES AND STATISTICAL MODELS

PARTITIONING	SYMBOLIC	MULTIVARIATE MODELS	UNIVARIATE MODELS
NONE	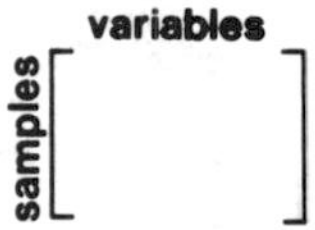 variables / samples	principal components analysis factor analysis cluster analysis	a priori restriction to one factor or component
VARIABLES	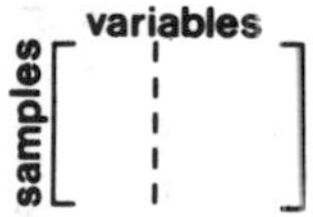 variables / samples	canonical correlation analysis	multiple regression and correlation
SAMPLES	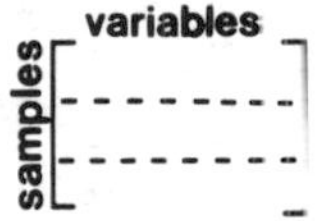 variables / samples	MV analysis of variance and discriminant analysis	analysis of variance
VARIABLES AND SAMPLES	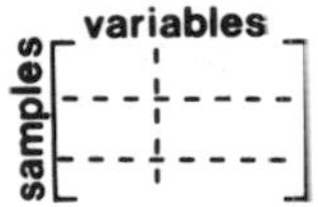 variables / samples	MV analysis of covariance and discriminant analysis with covariance	analysis of covariance
MULTI-DIMENSIONAL	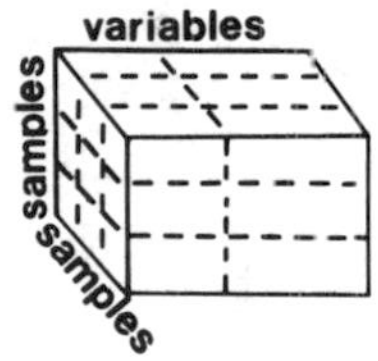 variables / samples / samples	factorial MV analysis of variance and covariance designs	factorial analysis of variance and covariance

FIGURE 3.3 Data matrices and statistical models

must be known so that a sampling design appropriate to tests of hypotheses can be formulated. Otherwise one is conducting a monitoring study to detect impact, rather than an impact study to test against the null hypothesis of no change due to impact. Third, it must be possible to obtain measurements on all relevant biological and environmental variables in association with the individual samples. Measurements for an area covering a number of samples may be adequate for description, but they are useless for hypothesis- testing. Fourth, an area that will not receive the impact must be available to serve as a control.

The first and fourth of these prerequisites imply that controls in both space and time are necessary. This is so because evidence for impact effects on the biological community must be based on changes in the impact area that did not occur in the control area (see Section 2.3.4). If the spatial control is missing and only before- and after-impact samples from an impacted area are available one runs the risk that a significant change may be unrelated to the impact. The change might have occurred anyway. Sudden lethal environmental changes can occur without warning (for an example see Dahlburg and Smith 1970) and sometimes without their being obvious. If the temporal control is missing, one may not detect that a difference between an area subject to the impact and an area not subject to it existed before the impact occurred. For example, when high mercury levels were found in some lakes in remote and uninhabited drainages it was a serious blow to the hope that particular instances of mercury pollution could be conclusively related to specific industrial sources. A firm accused of mercury pollution need only say: "It might have been like that before we started operation."

With reference to Figures 3.1 and 3.2, the four prerequisites define a design with at least one time of sampling before and at least one after the impact begins, at least two locations differing in degree of impact, and measurements on an environmental as well as a biological variable set in association with each other. This design corresponds to those in all combinations of rows 3 and 4 with columns 3 to 6 of Figure 3.2.

An optimal impact study design is therefore necessarily an areas-by-times factorial design in which the evidence for impact effects is a significant areas-by-times interaction (a data matrix and statistical model of the type shown in the fifth row of Figure 3.3). This is analysis of variance (or ANOVA) terminology, and indeed ANOVA represents an excellent statistical analysis approach in this situation (Sections 2.3.9 and 4.1). However, thoughtful scientists have always chosen such a design as the logical one for obtaining unambiguous results, even in studies where no statistical analysis is planned. An elegant example is provided by Barnett (1971) who contrasts the effects of a heated effluent on coastal fauna to before-impact data, with a control station three kilometres away. Results are presented in effective graphical and tabular formats, but with almost no statistics. Another example is provided by Hendricks et al (1974), who evaluate the effect of a paper mill effluent on bottom fauna in an estuarine environment using samples taken before and after- impact and above and below the effluent. Although both authors fail to follow through with optimal statistical analysis methods, the cases are good examples of optimal impact study design. Unfortunately scientists in a hurry often do not use such a design and sometimes attempt to cover up that fact by executing statistical dances of amazing complexity around their untestable results.

Given that the four prerequisites for optimal impact study design are satisfied, choice of a particular sampling design and statistical methodology should be based on the following criteria. First, one must be able to test the null hypothesis that any change in the biological community of the impact area, over a time period which includes the impact, does not differ from a control area. Second, it must be possible to relate to the impact any demonstrated change unique to the impact area and to separate effects caused by natural environmental variation unrelated to the impact. Third, the analysis method must lead to an efficient and effective visual display of both change due to impact in relation to other sources of variation and the relationship between impact-related change in the biological variables and impact-related change in the environmental variables. Fourth, it must be possible to use the results for subsequent biological monitoring to detect future impacts of the same type. Fifth, the test of the null hypothesis that no change due to impact occurred must be as conservative, powerful, and robust as possible (Section 2.1.6).

3.3 OPTIMAL AND SUBOPTIMAL DESIGNS—THE MAIN SEQUENCES

The many possible designs for sampling and statistical analysis in environmental studies can be classified into five general categories on the basis of the prerequisites for an optimal design. The sequential procedure for deciding to which category an environmental study belongs is outlined in Figure 3.4 in the format of a key. Each category is called a main sequence because it suggests a set of possible methods (and excludes others) in a methodological sequence proceeding from choice of sampling design and data transformations to hypothesis -testing, descriptive statistical analysis, and effective formats for presentation of the results. In Section 4 each

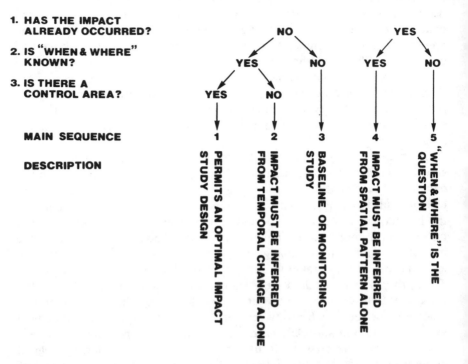

FIGURE 3.4 The decision key to the ''main sequence'' categories of environmental studies.

main sequence and the decisions that determine the choice of options within it are considered in turn.

Figure 3.4 indicates that the answers to three questions determine to which of the five main sequences a study belongs. If the impact has not yet occurred, before-impact (or baseline) data can be collected. All too often, however, the environmental biologist is first called in after the effects of an impact have become obvious. If the impact has not yet occurred, the next question is whether the location and time of occurrence of the impact are known. If they are known, the type of impact would also be known. A control in time exists by definition, and when a spatial control (an area not subject to the impact) exists, we arrive at main sequence 1. With measurements available on relevant biological and environmental variables in association with individual samples, this permits an optimal impact study design. If the spatial control is missing, the impact effects must be inferred from observed changes and it must be assumed that the observed changes would not have occurred if the impacted area had not been impacted. This leads us to main sequence 2. When the impact has not yet occurred and it is not known when or where it will occur, then a monitoring study is appropriate, which is main sequence 3. That is, we wish to detect an impact when and if it occurs by monitoring for its effects. As previously indicated, the nature of the impact may or may not be known. If we wished to assess the effects of an expected impact and did not know when or where it would occur but could recognize it if it did, then we would have a special case of either main sequence 1 or main sequence 2, where the duration of the before-impact portion of the study would be unknown.

Finally, if the impact has already occurred, the type of impact and its time and location of occurrence are probably known as well. Unless preexisting baseline data are available, impact effects must be inferred from spatial differences (among areas differing in degree of impact). This leads to main sequence 4. To the extent that preexisting baseline data can be used, we return to main sequence 2. If an impact has occurred and its nature is known, but the time and place of its first occurrence can only be hypothesized, then we have main sequence 5. The mercury pollution problem is an example.

Within each main sequence a number of decisions must be made, and for the purpose of discussion here they are allocated to five general decision areas. First (Sections 3.4 to 3.7), how many and what kinds of variables should be used and how should they be derived, coded, or possibly converted to another form? Second (Section 3.8), what special considerations influence choice of sample unit size and replicate sample number? Third (Section 3.9), what procedures permit preliminary screening of data for aberrant outliers, failures of assumptions, and so on? Fourth (Section 3.10), what computer programs and packages are available to provide the appropriate statistical analyses? Fifth (Section 3.11), what are appropriate visual display formats for the results?

3.4 FORMS OF SIMPLE VARIABLES

3.4.1 Quantitative, binary, and other category data

Little need be said about quantitative data—the most commonly used form that is generally considered to be the best and most informative type in some sense. Much of the discussion of Sections 2.3 and 3.4 to 3.11

deals with properties of quantitative data and continuous variables, and the problems encountered in their use. The emphasis here is on binary and other category data, including discussion of their advantages as well as disadvantages compared with quantitative data.

Good general references on data types, with emphasis on multivariable applications, are Anderberg (1973), Marriott (1974), and Clifford and Stephenson (1975). Chapter 5 in Grieg-Smith (1964) provides an excellent summary of simple statistical methods for analysis of relationships between biological criterion variables and environmental predictor variables, with various combinations of quantitative and qualitative data. Coefficients of association or similarity between different samples or different species based on quantitative and qualitative data are extensively reviewed in books by Anderberg (1973) and Orloci (1975a), and also by Krzanowski (1971), Orloci (1972), and Goodall (1973a). For testing significance of association with binary data see Cole (1949, 1957b) and Goodall (1967). Pielou (1969) shows how presence-absence species data, as number of species *not* in common in sample pairs, can be used in a principal coordinates analysis either as an ordination in its own right or for conversion to quantitative data for further analysis (Section 3.4.3). A sequential procedure for testing against $H_0: p = 0.5$ versus $H_A: p \neq 0.5$ in binomial trials was presented in Section 2.3.9.

It is conventional wisdom that binary or other category data are inferior to quantitative data—say, species presence-absence compared with species abundances. Certainly one speaks of percentage of information retained when considering conversion of quantitative data to category or binary form, but this is information in the technical sense equivalent to variation and not necessarily in the sense of ecologically meaningful. The amount of information (= variation) carried by the quantitative part of the data can be measured (Section 3.4.3) but often more of the quantitative than of the binary component is noise, because presences and absences tend to be more robust to the inevitable errors and biases in sampling (Section 2.3.9). Keup et al (1966) argue that changes in species dominance are important in biological assessment of impact and therefore that presence-absence data are less sensitive to impact effects. I disagree. My own experience is that properly designed sampling for binary data does not yield inferior results in comparison with quantitative data. Ecologically meaningful information obtained *per unit of cost or effort* may be greater with the use of a well-planned sampling and statistical analysis sequence for presence-absence data. It is misleading to think only of the results that would be obtained from dichotomizing quantitative data *after* they have been collected. This can be appropriate (see Section 3.4.3), but it does not provide the opportunity to choose a sampling method, sample unit size, and sample number most appropriate for presence-absence data (Section 2.3.9) and to reduce the cost of sample processing by not having to count all individual organisms. With an appropriate sample unit size, frequency of occurrence can be as sensitive a measure of species dominance as mean density. Noy-Meir et al (1970) vary quadrat size in presence-absence sampling of desert vegetation and find that small quadrats work most effectively (see also Sections 2.3.8 and 3.8), with larger quadrats tending to pick up "stragglers" outside their main distribution.

Lambert and Dale (1964) argue that "whereas presence-or-absence records form a self-contained logical system, all quantitative data are truncated in that no record is possible except zero for the absence of a species: although the amount by which a species is present can be recorded, there is no corresponding measure of the extent to which it is absent." They feel that meaningful qualitative pattern can often be obscured with quantitative data. The ambiguity of species absence, which was discussed in Section 2.1.9, applies to both quantitative and presence-absence data. Even a reduction in quantitative abundance is more ambiguous than is increased abundance—a species generally has one multivariate environmental optimum, but many possible suboptimal environmental variable combinations. Reduction of species abundance *may* be an appropriate measure of impact, perhaps even a better measure than increased abundance of pollution indicator species as suggested by Keup et al (1966), but this would be true only when there is a temporal control to establish the status of the species before impact.

Hurlbert (1969) uses simulated data to show that species association coefficients based on abundance are easily influenced by extraneous factors and concludes that such measures of association are less ambiguous if based on presence-absence data. Peterson (1976) studies the distributions of species of living and dead mollusks in two California lagoons and finds that species presences and absences are generally more reliable and ecologically interpretable than relative abundances. Allen (1971) reaches similar conclusions in his study of algae on terrestrial rock surfaces in Wales. There is no universal answer, the choice of data type depending very much on what you want. Smartt et al (1974) investigate the properties of vegetational data using seven data types, from the same field sampling, in a variety of statistical analyses including correlation analysis, hierarchical cluster analysis, and principal components analysis. They conclude that results based on qualitative data emphasize species richness and diversity and are best used where the interest is in species-environment relationships. If estimation of biomass or production is the primary interest, quantitative data will better reflect such differences.

Binary data can often be used, with good results, in multivariate statistical procedures intended for quantitative data. Buzas (1972), for example, uses presence-absence data in canonical variate analysis (= multiple discriminant analysis) and finds that the results are better in some ways than with the data as proportions or densities. See Section 3.4.3 for other examples and discussion.

There exists a more varied repertoire of statistical methods designed for binary and other category data than most ecologists realize. See Marriott (1974) for a brief general review of multivariate procedures for binary data. Linear logistic models for regression where the criterion variable(s) are qualitative can also be used for other statistical analyses, such as multidimensional contingency tables (Section 2.3.9) and two-group discriminant analysis. This approach, as described by Cox (1970), unifies methods that "concern the analysis of . . . data in which an observation takes one of two possible forms, e.g. success or failure. The central problem is to study how the probability of success depends on explanatory variables and groupings of the material." These models express a logistic transformation of probability of success (of a value of 1 as opposed to 0) as a function of a linear combination of unknown parameters. For computer programs see Lee (1974).

For multidimensional contingency table analysis of ecological data, with examples, see Fienberg (1970). Pielou and Pielou (1967) and Pielou (1972) present a somewhat different approach to the same analysis problem. Although most ecologists are aware that "all possible t-tests" is not a valid substitute for a properly designed ANOVA when analyzing quantitative data, relatively few realize that several two-way contingency table analyses are not a valid approach to analysis of multidimensional contingency table data. One reason, of course, is that if several analyses are done at the $\alpha = 0.05$ level, the conclusions based on all of them will not be at that Type I error level. Another is that an oft-ignored assumption in any two-way table analysis is a homogeneous population within a cell. Cox (1970) points out that the main problem in such two-way analyses is that other factors often influence the response, that is, "non-constancy of the probability of success within groups." This means that it is not even proper to do *one* two-way contingency table analysis if there are additional factors that are relevant to the frequencies within the cells of the two-way table. The additional factors cannot be ignored, but should be included as additional dimensions in the contingency table even when their influence is not the subject of interest. Cox provides an example showing how failure to do this can produce very misleading results. As with quantitative multivariate analyses, the problems of interpretation of the results increase with the number of variables (Section 3.6). The unknown functions of the predictor variables (the dimensions) are usually arranged as a hierarchical series of hypotheses (see Section 2.1.2). Here it is less easy to pretend that one is doing meaningful hypothesis testing when actually engaged in a sloppily conceived fishing expedition than it is, say, with multiple regression on quantitative variables. This could explain the much greater popularity of multiple regression. Fienberg (1970) believes that in most ecological analyses a two-factor model is adequate. When the analysis is multivariate in the strict sense of > 1 response (or criterion) binary variable, Cox (1972a) indicates that two approaches are in common use: use of 0/1 data in quantitative multivariate statistical analyses (see above and Section 3.4.3) or use of multidimensional contingency table methods despite the "one response variable" property of the models. Other options are considered in a clear nontechnical presentation.

What is the efficiency of logistic regression models compared with their quantitative variable counterparts? If we look at discriminant analysis (see Section 4.4) as one example, Cox (1972b) provides a model for discrimination between survival and death as a function of time and an explanatory environmental variable, using the linear logistic model. Cox and Brandwood (1959) construct a more sophisticated model in a fascinating problem of discrimination between the earlier and later works of Plato. Efron (1975) examines the efficiency of logistic regression as developed by Cox, and concludes that it tends to be approximately one-half to one-third as effective as normal discriminant analysis.

An interesting statistical model of structure in binary data is Latent Structure Analysis (Lazarsfelt and Henry 1968), which is discussed by Cox (1972a) and Marriott (1974). It assumes that "binary variates may be related . . . to assumed binary factors [which] can be regarded as defining a grouping of the observations" (Marriott 1974). However, the model also assumes that *all* relationships among the binary variables are caused by the underlying group structure. If the data were species presences and absences, this would be equivalent to assuming that species distributions *within* environmentally determined species assemblage groups (or faunal areas) were independent, and not correlated because of biotic interactions or microenvironmental patchiness. Most ecologists would agree that this is not a reasonable assumption. Also, there should be a "strong prior expectation that the classes have a clear physical existence" (Cox 1972a), which would mean that faunal areas must have different and internally homogeneous species abundance compositions and sharp boundaries. A good approach to ordination and clustering with binary, or mixed, data is principal coordinates clustering (Lefkovitch 1976), as described and illustrated in Section 3.4.3. A simple approach based on chi-square, analogous to factor analysis and appropriate for definition of indicator species (see Section 3.6), is association analysis (Williams and Lambert 1959, 1960). MacNaughton-Smith (1963) describes a procedure for clustering binary data so that the clusters formed are the best predictors of (differ maximally on) some criterion variable.

A variety of nonparametric methods (Sections 2.1.7, 2.3.9) are appropriate for qualitative data. Nonmetric multidimensional scaling is an ordination procedure based on rank distances (see Section 3.4.2) which can be used with binary data (e.g., Fisher 1968), as can the nonparametric analogue to canonical variate analysis described by Mantel (1970). Hypotheses having to do with binary or other category data in sequences of samples can also be tested. For example, Cochran's (1950) Q-test, which is described in Siegel (1956), could be used to test whether a species list (e.g., Audubon bird counts) changes over a series of years (see Section 4.2). Knight (1974) and Pielou (1975) discuss and illustrate tests of hypotheses about species relationships based on samples from transects across environmental gradients, using presence-absence data.

3.4.2 Ranks

Ranks can be used as a transformation of quantitative data intended for subsequent analysis by standard statistical methods, as discussed and illustrated with an example in Section 2.3.9. Conversion of large data sets to ranks is done easily in computer packages such as SAS (Service 1972, Barr et al 1976), and the rank values can then be directly entered into any desired statistical analysis procedure. Ellis (1968) describes programs that provide ranking summaries of quantitative biological data.

Use of ranks in standard statistical analysis procedures has not been common. Rounsefell and Dragovitch (1966) use rank abundance of the red-tide organism *Gymnodinum* as the criterion variable in a nonlinear multiple regression model. Woodward and Overall (1976) use ranks in a principal components analysis, in a comparative study of several ordination options. Gower (1967a) shows that the results of principal components analysis on ranks can be displayed on the surface of a sphere or hypersphere and that map-type projections (e.g., Mercator, conic) can therefore be used. As for cluster analysis methods, "ranks have rarely been used; they deserve more attention, for subjective rankings of species can be observed quickly and perhaps more reliably . . ." (Goodall 1973b).

Simple nonparametric statistical analyses based on ranks are well known (see Siegel 1956, Mosteller and Rourke 1973). Rank correlation, for example, may be used whenever there is likelihood of severe nonlinearity in a bivariate relationship. Levandowsky (1972) and Sprules (1977) use rank correlation to analyze relationships between scores on principal component axes (Section 3.4.3), derived from species abundance data, and environmental variable values. It should be remembered that the nonlinearity cannot be too extreme—such methods assume that the relationships are at least monotonic (ascending or descending, with no maxima or minima).

A nonlinear ordination procedure called nonmetric multidimensional scaling (MDS), which is based on ranked distances between samples in the "variable space," is rapidly gaining in popularity with ecologists. Press (1972) and Marriott (1974) provide brief descriptions, and the two early papers by Kruskal (1964a, b) are still worth reading. Computer programs are now widely available (Rohlf et al 1974, Gauch 1976, Kruskal 1977). MDS is similar to principal coordinates analysis (Section 3.4.3) in that the starting point is an n-by-n matrix of similarity measures between samples, and in both there is great flexibility of the choice of the similarity measure. With MDS, "given only rank order of pairwise resemblance between objects [it is possible] to imbed them in a euclidean space such that the rank order of pairwise distances preserves (or minimally distorts) the given rank order of resemblances" (Friedman and Rubin 1967). In a zoogeography application of nonmetric MDS Whittington and Hughes (1972) conclude that it "gives the best representation of n related objects in a given number of dimensions such that the inter-point distances correspond as nearly as possible to the relationships between the objects." Figure 3.5 is taken from Rohlf (1970) and illustrates the amount of distortion introduced by methods of summarizing phenetic relationships among n species. There are two clustering methods (a and b), principal components analysis (c), and nonmetric MDS (d). Note that the last shows the least scatter and therefore represents the best fit. Fasham (1977) compares various ordination techniques including MDS in applications to ecological data, as does A. J. Anderson (1971) in an earlier paper. There are two disadvantages to MDS as well as an inadequacy common to most ordination methods. It is a computationally demanding ordination procedure, and the dimensionality of the solution may be difficult to determine. Dimensionality is commonly assessed by examining results produced by solutions for different numbers of dimensions, or it is estimated from the minimum spanning tree (see Section 3.9), and the ecological interpretation of the results may depend on this uncertain dimensionality. Nonmetric MDS has the defect of most ordination procedures, and of all statistical methods based on ranks, that nonlinear relationships must be at least monotonic or a spurious dimension will be produced. Species abundance-environment relationships, for example, are often not monotonic (see Sections 2.3.9, 4.4) and even MDS fails to adequately describe them (e.g., Fasham 1977). Whether nonmetric MDS ordination is superior

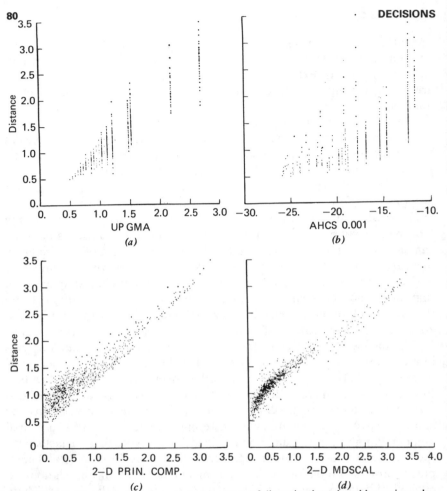

FIGURE 3.5 Comparisons of amount and pattern of distortion introduced by various clustering and ordination procedures. The ordinates are average distances between pairs of species and the abscissas are estimated distances from the various methods. Reproduced with permission from Figure 9 of Rohlf (1970).

to, say, principal coordinates analysis used with appropriate similarity coefficients, to an extent that the greater computational demands are justified, is uncertain.

Freeman et al (1967) give tables "for determining the number of observations to be made on $k = 3, 4, 5$ normal populations of common unknown variance, if the k population means are to be correctly ranked with high confidence." Freeman and Kuzmack (1972) extend this to $k = 30$. The minimum difference between means and the acceptable probability α of the Type I error must be specified. If preliminary sampling established approximate among-location differences in mean values of some impact-predictor variable, say the abundance of an indicator species, locations covering an area that includes a pollution source could be ranked according to degree of impact with some specified confidence $(1 - \alpha)$ that the ranking is correct.

3.4.3   Interconversion of data types

The conversion of data from one type to another can be viewed as data transformation in the broad sense (Section 2.3.9). A data set of mixed types is usually converted to one type, unless the planned statistical analysis method accepts mixed data as input. Sparse data matrices (with many zeros) can be considered to be mixed data, since they usually should be converted to a more condensed form, which could be any one of several data types. Even when data are all of one type it is sometimes desirable to convert them to another type, either because the information of interest will be better conveyed (see Section 3.4) or because the planned statistical analysis method assumes a particular data type.

The conversion of binary and other nonquantitative data to quantitative form is easily done. Principal components analysis (PCA) transforms observations on a set of p variables $X_{j=1,p}$ (which may or may not be quantitative) to observations on a new set of quantitative variables $Y_{i=1,p}$ where the new set has the following properties (see Marriott 1974):

1. Each new variable Y_i is a linear additive function of the old variables: $Y_i = b_{i1}X_1 + b_{i2}X_2 + \ldots + b_{ip}X_p$.
2 The first new variable extracted (Y_1) is that linear additive function which accounts for the largest possible amount of variation in the data.
3. Each additional new variable $Y_{i=2,p}$ accounts for the largest possible amount of *remaining* variation, independent of (= orthogonal to) the previously derived principal components (PCs) $Y_{i=1,2,\ldots}$. Therefore the new variables Y_i will be uncorrelated (= independent), unlike the original variables X_j. They will also tend to be more normally distributed, because of the central limit theorem.

See Section 2.3.8 regarding the consequences of the central limit theorem, and Section 2.3.9 where use of PCA as a transformation technique was mentioned. Use of PCA for reduction of variable number is considered in Section 3.6, and for avoiding ratios and dealing with intractable multivariate data in Section 3.5.3. Harris (1975) discusses use of PCA for transformation to uncorrelated variables for subsequent analysis.

PCA is most easily grasped in geometric terms (see Gower 1967a), as illustrated in Figure 3.6 for two variables. The first PC Y_1 is that straight line axis which follows the major trend in the data. With only two variables

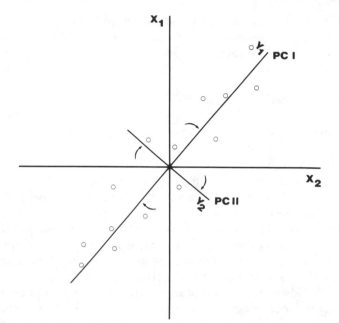

FIGURE 3.6 A diagrammatic representation of principal components analysis (PCA), showing how the principal components $Y_{i=1,2}$ are related to the original variables $X_{j=1,2}$.

(X_1 and X_2) the second PC axis Y_2 is then defined as the axis at right angles to Y_1. With more variables the second PC Y_2 would be obtained by rotating an axis orthogonal (= at right angles) to Y_1 until it was lined up with the largest *remaining* variation, and so on, until all new axes $Y_{i=1,p}$ had been found.

Principal coordinates analysis (Gower 1966a) is closely related to principal components analysis. In the latter the "n samples by p variables" data set is converted to a p-by-p covariance or correlation matrix. (Use of the correlation matrix is the equivalent of using data standardized to unit standard deviation on each variable. This is often not desirable with biological data. See Sections 2.3.9 and 3.9.). The roots and vectors (sometimes called the eigenvalues and eigenvectors) of this p-by-p matrix are then found, and correspond respectively to the variance associated with each PC, and to the PC coefficients b_{ij} in the function $Y_i = b_{i1}X_1 + b_{i2}X_2 + \ldots$. The b_{ij} are usually scaled so that the sum of their squared values add to 1. In principal *coordinates* analysis the n-by-p data set is instead converted to an n-by-n matrix of sample similarity measures, and then the roots and vectors of this matrix are found. The analysis proceeds as before. If the similarity measure is appropriately defined, the solution is identical to the PCA solution. The similarity measure can be defined appropriate to any type of data so that "with this representation there is no need for all (or any) of the variates to be quantitative" (Gower 1967b). However, Lefkovitch (1976, Appendix I) shows that all the usual data types can be transformed so that a PCA on the p-by-p covariance matrix is also possible, and therefore the criterion for choosing between the principal components and the principal coordinates approaches should usually be the relative magnitudes of p and n. If $p >> n$ and p is large, principal coordinates would be the easier approach. Pielou (1969) clearly describes both analyses. Orloci (1975) discusses these and related methods, and provides computer programs in the BASIC language. The NT-SYS package (Rohlf et al 1974) offers programs for both analyses. Any computer program that finds roots and vectors of a symmetric matrix and allows input of that matrix directly, possibly as an option in addition to the usual procedure of starting from the n-by-p data matrix—such as SPSS (Nie et al 1975)—can be used for either analysis.

PCA should not be confused with Factor Analysis (FA). They are *not* the same thing (see Pimentel 1978), although PCA is often used as a first step in FA solutions. PCA is an explicitly defined mathematical procedure, as unambiguous as addition or $Z = \log_e X$, and it produces a new data set that is nothing more than a transformation of the old one (retaining all the information). As with any transformation, it is simple to go from observations on the new variables Y_i back to observations on the old variables X_j (by a matrix multiplication), and no information is lost in doing so (if none of the p principal components were discarded). On the other hand, FA is a collection of methods that is about as well-defined as cancer or the common cold. It is a variety of techniques for modeling intervariable correlation structure, and the statement that someone did factor analysis is not very informative. It is not generally possible to convert FA scores back to the original data. For discussion on PCA and FA, their use, similarities, and differences, see Marriott (1974), Harris (1975), and Pimentel (1978).

The paper by Lefkovitch (1976) is a particularly useful reference which describes how one can proceed from data of several possible variable types, including mixed data, through principal components or coordinates analysis to ordination, clustering, or any other statistical analysis using the PC scores as the new uncorrelated quantitative (or qualitative) variables. An example is worked out here, based on a sparse binary (species presence-absence) data matrix of dimensions $n = 96$ by $p = 25$ (Table 3.1). Twenty-five species were obtained from five tidal levels in two replicate core samples collected at each of 10 times over a 24-hour cycle. Four samples that were lost are ignored for the purposes of this example.

Table 3.1 A sparse binary data matrix for presences and absences of 25 species at 96 location-times from the intertidal zone near Bamfield, British Columbia

Time Code	Level Code															Species									
		1	2	3	4	5	6	7	8	9	10	11	12	13	14	15	16	17	18	19	20	21	22	23	24	25
1	1	1																					1			
1	1	1																								
1	2						1																	1		1
1	2																							1		
1	3						1																	1		
1	3																							1		
1	4										1				1									1		
1	4								1		1					1										
1	5					1						1				1						1				
1	5					1						1	1									1				
2	1			1																	1					
2	1											1	1										1			
2	2							1																1		
2	2																							1		
2	3							1																1		
2	3																							1		
2	4									1										1		1				
2	4									1								1		1	1	1				
2	5																		1	1	1	1			1	
2	5			1	1						1	1						1		1	1	1				

84

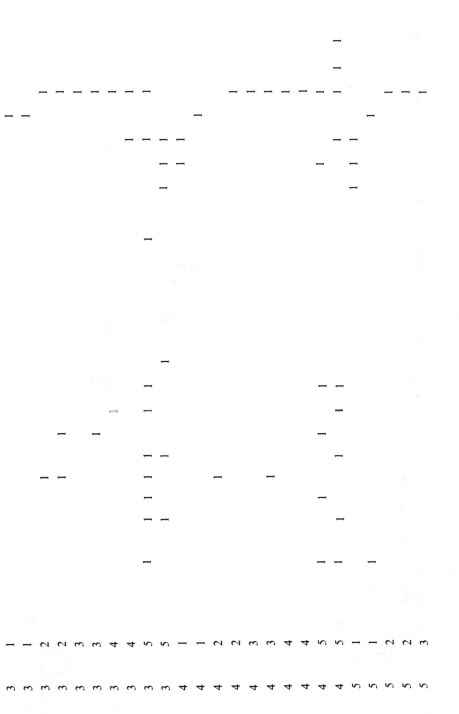

Table 3.1 A sparse binary data matrix for presences and absences of 25 species at 96 location-times from the intertidal zone near Bamfield, British Columbia

Time Code	Tide Level Code	1	2	3	4	5	6	7	8	9	10	11	12	13	14	15	16	17	18	19	20	21	22	23	24	25
5	3																							1		
5	4		1																						1	
5	4							1		1	1											1				
5	5					1	1	1		1	1	1									1	1				
Missing																										
Missing																										
6	1																									
6	2																							1		
6	2			1																				1		
6	3													1										1		
6	3																					1				
6	4							1			1	1										1				
6	4										1										1	1				
6	5				1	1	1					1									1				1	
6	5				1	1	1	1				1			1						1	1				
6	5							1			1	1									1	1				
7	1							1			1															
7	1																									
7	2																							1		
7	2																							1		

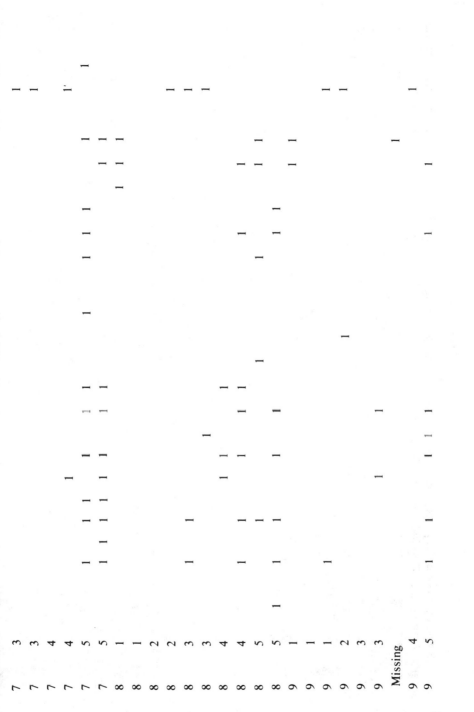

Table 3.1 A sparse binary data matrix for presences and absences of 25 species at 96 location-times from the intertidal zone near Bamfield, British Columbia

Time Code	Level Code																									Species
		1	2	3	4	5	6	7	8	9	10	11	12	13	14	15	16	17	18	19	20	21	22	23	24	25
Missing																										
10	1																							1			
10	1																							1	1		
10	2																								1		
10	2																								1		
10	3																								1		
10	3																										
10	4																								1		
10	4																								1		
10	5				1						1						1			1				1			
10	5		1							1	1									1			1	1			

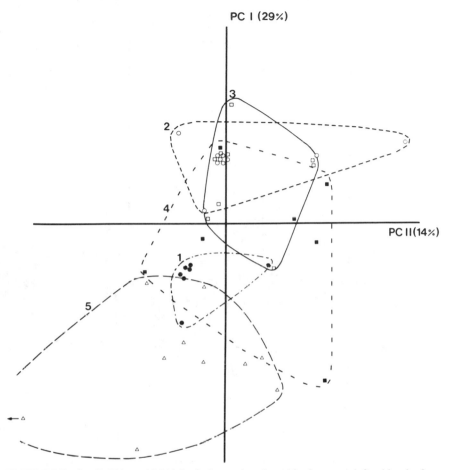

FIGURE 3.7   Bamfield intertidal biological samples plotted in the space defined by the first two principal components. Tide level 1 is the highest and tide level 5 lowest in elevation. See Figure 3.9 for tide levels and symbols.

Because $p < n$, the PCA approach is adopted. The data do not have a mean of zero, so a twenty-sixth dummy variable with a value of 1 for all samples is added, as described by Lefkovitch. The p-by-p = 26 × 26 matrix of cross-products is calculated, and the roots and vectors of that matrix are obtained. The first two principal components are shown as axes in Figure 3.7, and as the first two divisions in the corresponding hierarchical cluster analysis in Figure 3.8. This is an incomplete set of data, which will be augmented by at least as many species again, but even so it is apparent that at least the first PC carries meaningful information

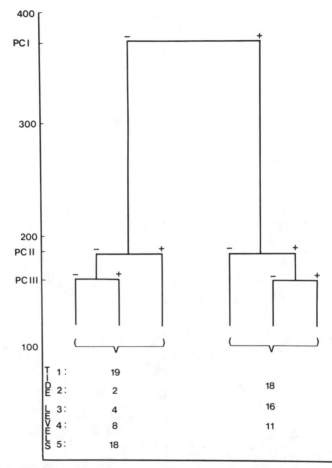

FIGURE 3.8 The hierarchical cluster analysis corresponding to the principal components analysis on the Bamfield biological data. The distribution of the 96 location-times is shown at the bottom. See the legend for Figure 3.7.

in that it defines a mid-tide level faunal zone. The PC I scores could be used as measures on a new variable Y_1 for any desired statistical analyses or graphical presentations.

Binary and mixed binary-quantitative data have been used in statistical methods designed for quantitative normally distributed data, often with satisfactory results. Obviously, a PCA can be carried out on binary data, as has just been shown (see also Blackith and Reyment 1971, Clifford and Stephenson 1975, and Thorpe 1976). Tests of significance would be risky for such drastically nonnormal data, but tests of significance are not often appropriate in PCA anyway (Marriott 1974). Glass et al (1972) conclude that ANOVA is robust even to the extreme nonnormality represented by dichotomous data, unless unequal sample sizes are involved (see Section 2.3.9). Gilbert (1968) uses simulated data to compare the performance of discriminant analysis on binary data with other analogous statistical analyses that are designed for binary data. She finds that "the loss involved . . . as opposed to any of the other procedures is too small to be of much importance" and concludes that the possibility of combining discrete and continuous variables makes the use of discriminant analysis in this way desirable. The conclusions of Moore (1973) are similar. Ivimey-Cook and Proctor (1967) use binary vegetation data in factor analysis with apparent success.

Category data, in either ordered or unordered categories, can also be converted to continuous variables using principal components or coordinates analysis (see Lefkovitch 1976), or they can be used in regression analysis or ANOVA by way of a dummy variable technique (Suits 1957). Gower (1967a) shows that proportion or percentage category data can be analyzed using PCA with particularly effective results. For example, consider the data in Table 3.2, which are for substrate composition in the same intertidal samples described earlier in this section. The data are expressed in terms of three percentage variables, but since their sums must always add up to a constant 100 percent it follows that all of the information could be expressed as values for only two variables. The results of a PCA are shown in Figure 3.9, which is a format appropriate to percentages-in-three-categories data. The dominant trend of variation (PC I) from fine to medium coarse, accounting for 92 percent of the variation, is obvious. Not only is an effective format for presentation possible, but the original correlated percentage category data is now converted to measures on two uncorrelated variables with increased normality of distribution and straightforward empirical interpretations.

What of the possibility of dichotomizing quantitative data (see Section 2.3.9), either because the information of interest is best carried by that type of variable (see Section 3.4.1) or because the planned statistical analysis requires it? Lefkovitch's (1970) procedure can be used to do this optimally, just as it can be used to convert data from binary to quantitative. With a sparse data matrix little information will probably be lost in converting all nonzero values to 1's (Williams and Dale 1964, Lance and Williams 1967a). Noy-Meir et al (1970) find that using "above some threshold" instead of "present" in a binary data cluster analysis gives improved results. (The threshold could be the median, as in Section 2.3.9). Norris and Barkham (1970) describe a field study where comparative analyses showed that quantitative data yielded little more useful information than

Table 3.2 Percentage substrate size fractions for samples from 96 location-times from the intertidal zone near Bamfield, British Columbia

Time Code	Tide Level Code	$<\frac{1}{4}$ in.	$\frac{1}{4}\,\frac{3}{4}$ in.	$>\frac{3}{4}$ in.	Time Code	Tide Level Code	$<\frac{1}{4}$ in.	$\frac{1}{4}\,\frac{3}{4}$ in.	$>\frac{3}{4}$ in.	Time Code	Tide Level Code	$<\frac{1}{4}$ in.	$\frac{1}{4}\,\frac{3}{4}$ in.	$>\frac{3}{4}$ in.
1	1	43	37	20	4	1	44	29	27	7	1	43	42	15
1	1	56	26	18	4	1	35	28	37	7	1	48	27	25
1	2	60	28	12	4	2	52	32	16	7	2	55	38	7
1	2	63	21	16	4	2	51	30	19	7	2	41	47	12
1	3	49	30	21	4	3	34	44	22	7	3	73	19	8
1	3	75	17	8	4	3	68	23	9	7	3	63	24	13
1	4	90	8	2	4	4	81	15	4	7	4	85	12	3
1	4	83	11	6	4	4	85	11	4	7	4	63	13	24
1	5	76	12	12	4	5	69	28	3	7	5	94	5	1
1	5	88	10	2	4	5	74	25	1	7	5	80	18	2
2	1	40	36	24	5	1	37	35	28	8	1	55	29	16
2	1	47	31	22	5	1	61	23	16	8	1	42	33	25
2	2	66	28	6	5	2	59	30	11	8	2	54	23	23
2	2	57	37	6	5	2	43	25	32	8	2	40	47	13
2	3	54	33	13	5	3	47	38	15	8	3	83	9	8
2	3	67	15	18	5	3	47	34	19	8	3	64	20	16
2	4	74	16	10	5	4	89	10	1	8	4	78	19	3
2	4	91	5	4	5	4	70	17	13	8	4	82	14	4
2	5	75	19	6	5	5	84	13	3	8	5	60	39	21
2	5	80	14	6	Missing					8	5	90	9	1

Var	Cat			
3	1	43	34	23
3	1	61	24	15
3	2	46	29	25
3	2	49	36	15
3	3	55	37	8
3	3	53	38	9
3	4	86	11	3
3	4	92	6	2
3	5	87	10	3
3	5	88	10	2
6	Missing			
6	1	51	23	26
6	2	59	31	10
6	2	63	16	21
6	3	83	12	5
6	3	70	15	15
6	4	90	7	3
6	4	80	18	2
6	5	73	25	2
6	5	83	15	2
9	1	50	36	14
9	1	35	31	34
9	2	57	33	10
9	2	52	31	17
9	3	73	20	7
9	3	86	9	5
9	Missing			
9	4	65	25	10
9	5	72	20	8
9	Missing			
10	1	38	36	26
10	1	29	40	31
10	2	50	33	17
10	2	48	31	21
10	3	73	18	9
10	3	74	18	8
10	4	76	15	9
10	4	89	6	5
10	5	74	24	2
10	5	87	12	1

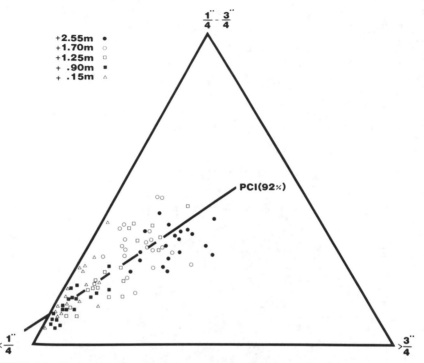

FIGURE 3.9 The results of a principal components analysis on data that are percentages in three substrate particle size categories, from the Bamfield intertidal study. Tide level locations are given as meters above mean low water.

did presence-absence data. Orloci (1975b) shows how to assess the relative information in the presence-absence and in the quantitative components of a set of data, allowing the decision on whether to dichotomize or not to be made with full knowledge. On the related subject of converting each quantitative variable in sparse data sets to two variables, one for the presence-absence component and one for the added quantitative component, see Williams and Dale (1962) and Orloci and Mukkattu (1973).

Finally, quantitative data can be converted to category data, for example, to render it appropriate for a contingency table analysis (e.g., Fienberg 1970). Cox (1957) presents a criterion for grouping designed to minimize loss of information resulting from the grouping, and also a useful tabulation of the information retained by various grouping schemes (different numbers of groups and distributions of samples among them) in comparison with the original quantitative data.

Clustering methods in particular are adaptable to data of any type or mixed types. See options in the CLUSTAN package (Wishart 1975) and papers by Goodall (1966b, c, 1973b), Lance and Williams (1967a,b, 1971), and Lance et al (1968). For use of mixed data in regression analyses or ANOVA see Suits (1957).

Computer packages and languages appropriate for carrying out conversions of data types are SAS (Service 1972, Barr et al 1976), APL (1969), and SPSS (Nie et al 1975).

3.5 DERIVED VARIABLES

Sokal and Rohlf (1973) note that most variables used by biologists are from "direct measurements or counts of biological material." Derived variables are "generally based on two or more independently measured variables whose relations are expressed in a certain way." Ratios, percentages, indices, and rates are all derived variables.

3.5.1 Indices

Sokal and Rohlf define an index as the ratio of one variable divided by another larger, standard one. However, it is used here in the broader sense of Pikul (1974) as "a mathematical combination of two or more parameters which has utility at least in an interpretive sense." Two types, species diversity indices and indices that are ratios of variables, are considered in separate sections (Sections 3.5.2 and 3.5.3) because of their importance.

As with any criterion or predictor variable it is important that index variables chosen be appropriate for the hypotheses and the statistical model that will be used to test those hypotheses (Sections 2.1.5, 2.3.1, and 3.6). Unfortunately, environmental biologists often formulate and use index variables intuitively, thereby avoiding clear thinking about hypotheses and intervariable relationships. There will be those who disagree, but I feel that arbitrary compounds of independent variables such as the compounds of density, frequency, and dominance used by West (1966) and Erman and Helm (1971) should be avoided. It is better to use individual criterion or predictor variables in linear additive models and then let the relative magnitudes and signs of the coefficients in the linear addive functions be the basis for interpretation of criterion or predictor variable relationships. Suitable transformation (usually logarithmic) of the variables will allow the interpretation of linear additive relationships among the transformed variables as multiplicative or ratio relationships among the original variables. See section 3.5.3 for further discussion.

For example, assume that we have an effluent-impacted area and a control area and we want an index of impact by that effluent. We might randomly sample in each area and use the log-transformed abundances of p species as variables (see Section 3.6 regarding variable selection) in a discriminant analysis (see Section 4.4 for discussion and an example). The linear additive discriminant function of the log-transformed species abundances would represent (by definition) the best predictor of that impact and the most efficient test of H_0: "no impact." If the design includes a temporal (before-impact) as well as a spatial control, we have a multivariate version of the single criterion variable example presented in Section 2.3.9 (see Section 4.1). Use of the index derived as the linear additive function from the spatial-by-temporal interaction term, as a predictor or indicator variable for biological monitoring, is illustrated in Section 4.3. Definition of impacted and unimpacted areas could be incorporated into the analysis by first performing a cluster analysis to group the samples into faunally homogeneous assemblages. If the two-group solution can be interpreted as impacted and nonimpacted groups of samples that are spatially and temporally contiguous, discriminant analysis could be used as described above to define an index of the faunal differences between those groups. If the groups are derived by cluster analysis, no significance tests would be appropriate, of course, since the groups were *created* so as to maximize differences on the discriminating variables. On the other hand if the groups are defined *a priori* (as effluent-impacted and control), tests of H_0: "no differences" in species composition between impacted and nonimpacted areas *would* be appropriate.

The use of derived indices as indicators of environmental quality in some general sense is not an approach about which I am enthusiastic (see also Section 3.5.2). Pikul (1974) discusses environmental indices of this kind and draws an analogy with economic indicators such as GNP. It is a good analogy, and the continuing disagreements over what such economic indicators mean at any given time and place should probably be kept in mind. Pikul presents an example of how to empirically evaluate and choose indices for applied use and lists over 100 indices that have been used in relation to environment. In a similar vein Maloney (1974) discusses the combination of "multiple attribute outcomes" into an overall index and also considers conversion of scales of measurement and graphical evaluation techniques.

3.5.2   Diversity indices

Pielou (1969) believes that the idea of species diversity was first introduced by Fisher et al (1943) in connection with the log-series distribution. The work of Margalef (1958a, b) popularized the concept among ecologists and created strong interest in information and entropy-based diversity indices. Over the next decade attempts to apply diversity to all communities, and thereby simplify and explain them, much resembled the attempt to explain everything in aquatic communities by pH during the 1920s (see Allee et al 1949). Just as the diversity approach was waning in theoretical ecology, Wilhm and Dorris (1968) introduced it to applied ecology with the argument that "These indexes express the relative importance of each species, are dimensionless, and are independent of sample size. Pollution results in depression of diversity . . . in the biotic community." Since then diversity indices have been extensively and often uncritically applied, without regard to the assumptions implicit in the various diversity formulae and the biases in their estimation and despite many published critiques and premature funerals. Standard references to information theory and to indices based on it, such as Kullback (1968) and Pielou (1969, Chapter 18), are often ignored. It is my own view that environmental studies have too much been sidetracked into the numerology of diversity index and related techniques. The literature has been filled with wrangling over which diversity index should be used of the countless ones proposed while there has been too little emphasis placed on the critical importance of proper sampling design, or on appropriate multivariate statistical analysis methods that can efficiently test exactly the hypotheses one wishes to test in environmental studies. At times the environmental biologist seems to resemble Eric Hoffer's (1951) True Believer: "When some part of the doctrine is relatively simple, there is a tendency among the faithful to complicate and obscure it. Simple words are made pregnant with meaning and made to look like symbols in a secret message. There is thus an illiterate air about the most literate true believer. He seems to use words as if he were ignorant of their true meaning. Hence, too, his taste for quibbling, hair-splitting and scholastic tortuousness."

Good general reviews of the commonly used diversity indices are by Cairns et al (1972), Fager (1972), Poole (1974), and British Columbia Lands, Forests and Water Resources (1974). A bibliography of "Biological indicators of environmental quality" by Thomas et al (1973) includes a section on diversity indices. Early plant ecology literature on diversity indices is in a bibliography by Goodall (1962). Dejong (1975) discusses linearizing transformations for several commonly used indices. Auclair (1971) provides what is probably the best example from field data (upland forests) of spatial and temporal patterns of variation, and covariation, of all the commonly used diversity indices. Some of the indices are defined and discussed below.

S: The number of species. Poole (1974) notes that it is the "only truly objective measure of diversity."

d: $(S - 1)/\log N$, where N is the number of individuals of all species. Any base logarithm could be used. See, for example, Margalef (1958a, b).

D: $1 - \sum_j [N_j (N_j - 1)]/[N (N - 1)]$, where N_j is the number of individuals of species j. Proposed by Simpson (1949), it is the probability that two randomly selected individuals will not belong to the same species (see Pielou 1969 and MacDonald 1969). Hurlbert (1971) and Smith and Grassle (1977) develop a generalization for the expected number of species in a random sample of m individuals, which for $m = 2$ is Simpson's Index D. The contribution by each species to this index is proportional to the probability of it appearing in a sample of m individuals. Therefore as m increases the contribution by rare species increases.

H: $1/N \log N!/(N_1!N_2! \dots N_S!)$, which is Brillouin's (1962) measure of the information content per symbol of a message made up of N symbols of S different kinds. It is appropriate as a measure of the uncertainty about which of S species an individual will belong to if it is randomly selected from a *fully censused* collection of N individuals.

H': $-\sum_j p_j \log p_j$, where p_j is the proportion of the population that is of the jth species. This is the Shannon and Weaver (1949) measure of the information content per symbol of a code which uses S kinds of discrete symbols with probabilities of occurrence p_j. Any base logarithm can be used for H or for H'. Whereas H is a measure appropriate to a finite fully censused collection, H' is defined only for an infinite population. The proportion N_j/N of species j in a random sample of N individuals from the population (which is all individuals of all species in the natural community) might be used as an estimate of p_j (e.g., as in Crossman and Cairns 1974). Unfortunately, $-\sum_j N_j/N \log N_j/N$ is a biased estimator of $-\Sigma p_j$ $\log p_j$ even *if* the natural community is really randomly sampled, which is unlikely (see Section 2.3.6). See Pielou (1969) for extensive discussion regarding these information-based indices. Pielou suggests that if the unknown infinite population is treated as a code, H can be used for a particular finite fully censused collection which is treated as a particular message using that code.

SR: The "rarefaction index" of Sanders (1968, 1969), which attempts to remedy the problem of the dependence of diversity estimates on sample size. In a plot of number of species versus number of individuals each collection is represented by the end point of a line which starts from the origin (zero individuals, zero species) and represents the numbers of species and individuals one supposedly *would* have obtained from samples of smaller sizes. Sanders assumed that each species would be represented in the reduced samples by the same percentage of individuals as in the original sample, but Fager (1972) and Simberloff (1972) use simulation studies (see Sections 2.1.7 and 2.3.9) to show that scaling down in this manner consistently overestimates the expected number of species. Heck et al (1975) present formulae for correct calculation of the expected number of species S_N in samples of sizes containing N individuals. Fager concludes that even with correct calculations the SR technique can be safely used for comparison of samples only under restrictive conditions which "seem biologically unrealistic."

Although diversity has been considered an intrinsic property of communities (Hairston 1964, McIntosh 1967), the recent view (Hurlbert 1971, Goodman 1975) is that diversity has reality only as a vague concept that combines two different and often independently varying (Sager and Hasler 1969, Tramer 1969, Moore 1975) components: number of species and equitability of abundance among them. Diversity indices are nothing more than multivariate variance measures of individual organisms across species. Poole (1974) comments that they are "Answers to which questions have not yet been found." Since this book strongly emphasizes the importance of knowing what question you are asking (Section 2.3.1), it correspondingly deemphasizes the use of diversity indices.

What of the supposed relationships between diversity and such community properties as information and stability? McIntosh (1967), for example, stated that "The concept of diversity is particularly important because it is commonly considered an attribute of a natural or organized community (Hairston 1964) or is related to important ecological processes." However, Goodman (1975) argues that the only meaningful way to define diversity is in terms of a particular index and that there is no meaningful definition by way of a relationship or analogy with community information, stability, or structure. Of the Shannon-Weaver index H' he comments that "There does not seem to be an ecological process that corresponds in any obvious way to this . . . [and] whatever the index does measure seems to have no direct biological interpretation . . . ." As for a relationship with community information or entropy, Pielou (1969) concludes that "these fashionable words have been bandied about out of their proper context . . . and have led to false analogies that produced no noticeable advance in ecological understanding." An example of the early arguments for a diversity-stability relationship is Pimentel (1961). Interest in such a relationship was stimulated by MacArthur's (1955) attempt to relate trophic complexity and community stability. However, there was much semantic confusion involved because of the lack of definition of "stability" (see Lewontin 1969). Paine (1969) and Engstrom-Heg (1970) argue that diversity and stability are not related in any simple way, and Goodman (1975) concludes that "the belief that more diverse communities are more stable is without support." Evidence to the contrary for terrestrial and marine, forest, and experimental terrestrial systems is provided by Buzas (1972), Auclair (1971), and Hurd et al (1971), respectively.

The use of diversity indices in applied studies must therefore be justified by their empirical value for specific purposes, rather than on theoretical grounds. In this regard there are serious problems in assuming that diversity as calculated from a sample is an unbiased estimate of the diversity of the community (see above), which may or may not be a serious problem when used for comparative purposes. See Section 2.3.6 for discussion on this point, as well as on the virtual impossibility of randomly sampling most natural communities. Bias in the estimators of the information-based indices is discussed by Lloyd and Ghelardi (1964), Pielou (1969), Bowman et al (1971), Goodman (1975), and Smith and Grassle (1977). Another source of bias is that "Diversity indices are frequently applied in the form of ratios of absolute diversity to the maximum diversity possible [and] the resultant indices can be shown to possess undesirable properties" (Peet 1975). The same point is made by Sheldon (1969). Although Fager (1972) recommends that diversity indices be relativized in this manner, Peet notes that "such scaling does not decrease, but much increases, the effect of sample size on the measurement." This is the same "ratios of variables make poor derived variables" problem that is discussed in Sections 2.3.8 and 3.5.3. Peet concludes: "The implication is that at best ecologists may have lost a fair amount of time calculating relatively meaningless numbers."

Another problem is that the assumption of a connection between high diversity and high environmental quality does not appear to be valid generally (e.g., Archibald 1972, Tramer and Rogers 1973, Livingston 1975,

Zimmerman and Livingston 1976). Diversity indices are not robust indicators in this regard because they depend on many factors other than environmental quality, such as seasonal and other temporal factors (Mackay and Kalff 1969, Holland and Polgar 1976, Menge and Sutherland 1976), latitudinal factors (Pianka 1966), trophic factors (Paine 1966), and spatial variation in natural environmental factors (Sheldon 1968, Recher 1969, Harman 1972, Hendricks et al 1974, Tietjen 1976). Also, contrary to Wilhm and Dorris (1968), diversity index estimates are *not* independent of sample size (McGowan and Fraundorf 1966, Peet 1975).

Diversity indices are attractive largely because they appear to reduce the information in large masses of data to single numbers. Information is lost in doing so of course, but that is to be expected. "Whichever index is used, it is bound to lose information just as a mean does because it is a summarization of data" (Fager 1972). However, one should ask whether a diversity index is the most efficient and biologically interpretable way to summarize biological data. There are certainly examples of their successful use, especially in environments where adequate baseline data have been obtained within particular geographic areas (e.g., streams and rivers; see Cairns et al 1972), or in studies that are themselves designed to have adequate spatial and temporal controls (e.g., Hendricks et al 1974). Successful applications to terrestrial flora, marine benthos, and marine phytoplankton are by Williams et al (1969), Rosenberg (1971, 1972), and Briand (1975), respectively. In all these studies the diversity index approach was in addition to other methods for analysis and display of the results, which were successful in their own right.

If species diversities are desired for comparative purposes, perhaps to supplement another method, simple indices such as S and d are biologically meaningful measures which are less ambiguous than—and often as informative as—more complex indices such as H and H'. Hurlbert (1971) makes this argument forcefully. In a simulation study Green (1977) found that S was a better indicator of biological change than was H'. Cairns et al (1971) appear to obtain as effective results by using S as an indicator in a Roanoke River study as by using their variant of H'. Harman (1972) and De March (1976) provide examples of successful use of S as a biological criterion variable. Poole (1974) argues for its use, and Barton and David (1959) show that for statistical treatment the binomial distribution adequately approximates that of S. Margalef (1958a, b), in his original papers which introduced information-based diversity indices to ecology, used d as well as H on the grounds that he found it to be just as good in practice and easier to compute. Rosenberg (1973) found that d was much *better* than H in showing biological patterns related to pollution.

Finally, the strongest argument against the use of diversity indices as

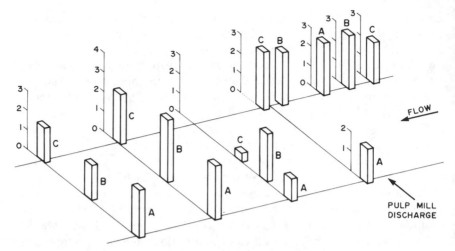

FIGURE 3.10 Levels of the diversity index H' for benthic fauna at stations in the Winnipeg River near a pulp mill effluent. The data are from research by Environment Canada, Freshwater Institute, Winnipeg.

derived criterion or predictor variables in environmental studies is that other statistical methods retain more of the information in the biological data while reducing them to a more useful and ecologically meaningful form. If it is necessary to have an index as the biological variable, then indices derived to be maximally efficient for specified purposes are best (see Sections 3.5.1 and 4.3). A multivariate statistical approach is often most appropriate, as for example in a study of benthic faunal patterns in the Winnipeg River related to a pulp mill effluent. Figure 3.10 shows the results of the diversity index (H') approach. Some depression of diversity values is apparent immediately below the effluent discharge, but beyond that the downstream values are about the same as the unpolluted upstream values. Figure 3.11 shows the results of a cluster analysis approach to the same data. Patterns related to degree of impact are obvious, and the upstream stations form a distinct group. The cluster analysis of course retains information about *which* taxa characterize each of the stations whereas the diversity indices do not. See Section 4.4 for a detailed presentation of a cluster analysis approach, which is then followed by a discriminant analysis on environmental predictor variables.

3.5.3 Ratios and other kinds of intractable data

Four kinds of data that cause difficulties in statistical analysis are considered: data sets with ratios as variables, data sets obtained from ob-

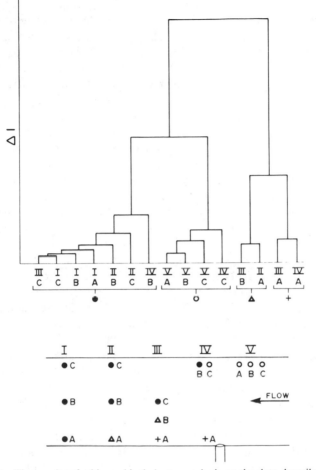

FIGURE 3.11   The results of a hierarchical cluster analysis on the data described in Figure 3.10. The ordinate ($\wedge I$) for the dendrogram is the gain in information on fusion of two stations or groups of stations.

servations on heterogeneous material (mixtures of distributions), data sets with missing observations, and problems specific to multivariate statistical analysis. Quantitative data with many zeros, forming sparse data matrices, are another kind of intractable data and were considered in Section 3.4.3.

Problems with ratios as derived variables were mentioned in Section 2.3.8, and in Section 3.5.2 with special reference to standardized diversity indices. Ratios are widely used in biology, especially in systematics and ecology where it is often considered desirable to standardize, or relativize, variables. An excellent recent reference on the subject is by Atchley et al (1976), who comment that "In spite of the widespread use of ratios and proportions, the vast majority of biologists are unaware of the statistical consequences resulting from compounding variables into ratios." See also the "Point of View" section in subsequent issues of *Systematic Zoology* for discussion provoked by the article. Although statisticians have been aware of the problem since the work of Pearson in 1897, few introductory statistics textbooks discuss it. An exception is Sokal and Rohlf (1973) who summarize the disadvantages of ratios as: (1) increased variability in comparison with that of the variables that were compounded into the ratio, (2) biased estimation of the true mean value of the ratio, (3) leading to unusual, nonnormal and possibly intractable distributions, and (4) tending to obscure rather than elucidate the intervariable relationships. Kendall and Stuart (1969) point out that creating ratios from normally distributed variables leads to the Cauchy distribution which has no moments or functions of moments.

Atchley et al use very large simulated data sets to examine the effects of compounding normally distributed variables into ratios. They consider correlations of three types: (1) ratios with a common denominator, such as $Y = X_1/X_2$ with $Z = X_3/X_2$, (2) a ratio with its own denominator, such as $Y = X_1/X_2$ with $Z = X_2$, and (3) a ratio with its own numerator, such as $Y = X_1/X_2$ with $Z = X_1$. They conclude: "The results . . . demonstrate that a number of large and systematic statistical changes occur with respect to the structure and underlying distribution of data when ratios are compounded between continuous variables." The worst situation is where there is high variability in the denominator of the ratio. Spurious correlations among uncorrelated variables are produced.

Ratios are often created by attempts at scaling, for example by dividing size of body part measures by an overall size measure in order to convert them to relative size measures. Ecologists commonly do the same thing with species abundances to express proportional abundance variation. Atchley et al show that the size-dependence is often *increased* by such an approach. Peet (1975) finds the same to be true of scaled diversity indices. He also shows that such relative diversity indices are extremely sensitive to small sampling variation, for example a change of species by one out of $n = 1000$ individuals producing a sizable change in the index value.

Proportions that are estimated as Y/n (or percentages as $100Y/n$) for differing numbers of samples n, and then used in the same statistical analysis, are another example of the poor behavior of ratio variables. Heterogeneity of variance is the probable result, and the arcsin square root transformation for proportions (Section 2.3.9) will not correct it—

the stabilized variance that results from proportion data so transformed is still a function of n (see Steel and Torrie 1960, p. 158).

Atchley et al also examine the effect of ratio variables on a multivariate analysis by performing a principal component analysis (see Section 3.4.3) on a simulated data set with $p = 6$ variables and $n = 1000$ samples. One of the variables (an overall size measure) was used as a scaling variable (dividing all the other variables by it) as is often recommended. They found that the analysis was greatly modified, that correlations were inflated and that the importance of principal components and the relative values of their coefficients were changed.

There is no need to use ratios as derived variables in most cases, because other analysis approaches will accomplish what ratios are supposed to and usually do not. Atchley et al point out that analysis of covariance often is the appropriate statistical model. Suppose, for example, that one wishes to determine whether, and in what manner, respiration rate in a particular species varies with temperature. Obviously, larger animals will tend to have higher oxygen consumption than smaller animals at any given temperature. Instead of creating a ratio variable by dividing respiration rate by weight for each animal and then, say, comparing log respiration rate per gram among different temperatures by ANOVA, it would be better to compare regressions of log respiration rate on log weight, by analysis of covariance, among the temperatures. The results of such an analysis are shown in Figure 3.12, taken from Green and Hobson (1970). Cochran (1957) is a basic reference to analysis of covariance and its assumptions. One assumption is that the covariate (log weight in the example) is fixed and measured without error, which will usually not be satisfied as it is not for log weight. However, Glass et al (1972) conclude that in practice this violation does not cause serious problems, although the power of the covariate to correct the ANOVA on the dependent variable tends to decrease and eventually go to zero as the variability of the covariate increases.

An alternative approach is to use principal components analysis (PCA). See Section 3.4.3 for discussion and examples. With both morphometric data and species abundances a log transformation, with no other standardization (see Section 2.3.9), is appropriate. The PCA, done on the covariance matrix (which is implied by ''no other standardization''), usually yields principal components of which the largest is an overall size (or overall abundance) component related to the original variables by coefficients all of the same sign and of magnitude roughly proportional to their size-abundance contribution (see Blackith and Reyment 1971, Marriott 1974). The other components represent *proportional* variation in size or abundance and have large coefficients with *opposite* sign for those original

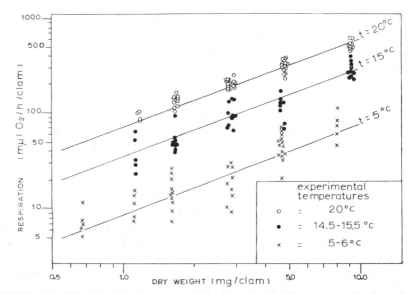

FIGURE 3.12 Respiration rate as a function of dry weight for the clam *Gemma gemma* at different experimental temperatures. Both axes are on logarithmic scales. Reproduced with permission from Figure 10 of Green and Hobson (1970).

variables whose proportional variation in size-abundance define the component. Since the original variables were log-transformed, a linear additive function of two variables, with opposite coefficient signs, can be interpreted as a ratio. If, for example, a component is a function of log-transformed water chemistry variables sampled over many water bodies and the largest coefficients are for calcium and total alkalinity, with opposite sign, then "variation in the proportion of calcium in the total dissolved salts" is a reasonable interpretation. If the first and largest component has coefficients all of the same sign and magnitudes roughly proportional to the contribution to total dissolved salts, it can be interpreted as a component describing variation in total dissolved salt load among lakes. If the variables are log species abundances and samples are from a grid covering an area where there is an effluent, a first component with coefficients all of the same sign will represent overall abundance variation over the area. A component with high coefficients of opposite sign for two species, and unusually high (or low) values of the component scores in the area just below the effluent, may be interpreted as a ratio of the abundances of two species which are good indicators of the pollution—one species that increases because of the pollution and one that decreases. See Section 3.5.1 for discussion on linear additive functions as derived variables. Kuris and Brody (1976) provide an example of the use of PCA in an ecological size-shape analysis.

It is not necessary that the components other than the first and largest overall size component be interpretable in a PCA carried out for this purpose. If proportional or ratio variation is the only variation of interest, the first component can be thrown out and all further statistical analyses of whatever kind performed on the other components. Either the PC scores can be used as new uncorrelated variables describing proportional variation, or the PCs other than the first can be used to back-calculate to the original variable values with the overall size component removed.

Finally, if ratios or proportions with variable denominators must be used as variables in statistical analyses, there *are* valid procedures for some analyses, which should be used. In particular, see Cochran (1963) on ratio estimates of proportions and related topics. An example is in Green (1968).

Sampling a mixture of distributions can easily arise without the biologist being aware of it, for example, if the taxon variable really includes more than one species and they have different spatial or temporal distributions. The two sexes or different year classes of a single species may differ in their environmental preferences or seasonal abundance variation. The most important thing is to always be aware of the possibility that it may be happening. If it is, it will be a source of variation that could obscure differences and make tests of hypotheses less efficient. Lachenbruch and Kupper (1973) use simulated data to examine the performance of a multivariate procedure (discriminant analysis) when a group that is assumed to have a multivariate normal distribution is in fact a mixture of normal distributions. They find that there is little effect on the results so long as the distributions in the mix do not differ too much. Statistically it is a problem of heterogeneity of variance caused by the mixture of distributions, and most statistical methods are fairly robust in the face of this violation (see Section 2.3.9). However, clear interpretation of results is not aided if more than one thing is going on under the label of what you think is a single homogeneous variable. Procedures for separating composite distributions into their individual component distributions exist and are commonly applied to size distributions based on several age classes. A graphical method for separating overlapping normal distributions (or distributions that can be normalized by transformation) is described and illustrated by Cassie (1954). An application of Cassie's method is in Green and Hobson (1970). The NORMIX procedure of Wolfe (1970) can be used to separate overlapping multivariate normal distributions (see Section 3.9).

Missing observations are of two kinds and the distinction between them is important. If they are logically missing, the analysis of the data must be appropriate for whatever unequal sample sizes or unbalanced design may result, and such analyses are often difficult to locate, to carry out, and to interpret afterward (see Section 2.3.3). If they are missing because of an unforseeable accident, it may be possible to estimate the missing values from the rest of the data. General references for problems with both kinds of missing observations are Steel and Torrie (1960), Snedecor and Cochran (1967), and Sokal and Rohlf (1969).

Suppose the null hypothesis is that a pollutant does not depress adult size in four species, as opposed to the alternative hypothesis that it does. Individuals of each species are to be collected in equal numbers and measured, from each of three locations (with pollutant concentrations of zero, medium, and high). If it is found that one of the species is totally absent at the location with the high concentration, observations for that species at that location are logically missing, and an unbalanced ANOVA design results. It is possible to test for significance of main effects in such a design, but tests for interaction are generally rendered ambiguous. Unfortunately the test of H_0: "no interaction" would be the test of interest in this example, corresponding to H_0: "no change in adult size with change in pollutant concentration." Woodward and Overall (1975) present an interesting discussion, with examples, of the use of multiple regression methods to perform multivariate ANOVA including nonorthogonal designs. Strategies for coping with nonorthogonality in univariate ANOVA are also discussed. Grizzle and Williams (1972) present linear logistic models for contingency table, or category, data which can handle problems where some cells are *a priori* zero as part of the model. Again, some interaction contrasts cannot be estimated, however. Recent trends of research work on analysis of incomplete multivariate data are reviewed by Rao (1972).

If observations are missing by accident, the first question is whether the observations are missing at random. Most methods for handling missing observations assume that they are. Even large data sets can be scanned, the incomplete data tabulated, and missing values evaluated for pattern. BMD computer programs BMD09D and BMD10D are examples of programs that do this (Dixon 1973). For univariate data missing observations can easily be estimated by analysis of covariance, and Coons (1957) describes the technique. With multivariate data missing observations are a common problem, and several authors have considered possible approaches. Elashoff and Afifi (1966) review the literature up to that date. Chan (1972) uses simulated data to evaluate various ways of treating missing values in discriminant analysis, including deletion of the incomplete samples, leaving blanks, substituting means, a regression method, and a PCA method. He concludes that either substitution of mean values or the PCA method are best. Frane (1976) discusses deletion of incomplete samples, substituting means, and several regression methods. Two regression methods are recommended, the choice depending on whether sample number n is large relative to variable number p, or vice versa. It is usually necessary to adjust degrees of freedom when testing hypotheses with data containing estimated values for missing observations (see Steel and Torrie 1960, Sokal and Rohlf 1969).

Multivariate statistical analyses with ecological data often break down because of intrinsic properties of the data. For the novice at using multivariate methods the reasons for such analysis breakdowns are often mysterious, but he need not throw up his hands and abandon use of the method. If dispersion matrices for one or more groups are singular (with zero determinants), inversion of those matrices or of pooled matrices based on them are impossible and many multivariate analyses cannot be carried out. It is even worse when the program proceeds, with excessive rounding error. Possible causes of singular matrices are perfectly correlated variables, all zero values for a variable, or an overdefined matrix (too many variables in relation to the number of degrees of freedom). Norris (1971) considers procedures for dealing with this problem. Deletion of problem variables is *not* recommended. A modified PCA transformation procedure is described and illustrated with an example. Lost information can be estimated.

3.5.4 Equation and test statistic parameters

Equation rate parameters are not often used as criterion variables, but they can be used very effectively. The parameters of growth models (see Section 3.7) can be used as descriptors of growth pattern to be related to environmental factors. Mortality rate, as estimated in laboratory toxicological studies using logit or probit models or from field data as percentage change in density over time (see Section 4.2), is another obviously useful variable for environmental studies. Often change rates of abundance (because of births or deaths) or of individual size of organisms are percentage rates, and the rate parameters in the models are equivalent to logarithmic transformations of those rates (see Section 2.3.9)—for example, the instantaneous mortality rate z in the model $N_t = N_0 e^{-zt}$ or the instantaneous growth rate K in the growth model $S_t = S_\infty (1 - e^{-Kt})$. Therefore, they are appropriately pretransformed for use as a variable in another statistical model.

Where rate of change of Y per unit time at different times is the required variable, we often have only data on the magnitude of Y at different times. The variable Y could be a biological variable such as species abundance, where we want rate of change in abundance, or an environmental variable such as total amount of effluent discharged at different times, where we want discharge rate. Curve fitting procedures such as polynomial regression could be used to fit the time trend $Y = f(t)$, and then the slope of the fitted model at any time t could be easily determined (as the first derivative dY/dt) and used as the estimate of the rate of change per unit time at time t. If Y were a species abundance, as in the first example, the first derivative of a polynomial function log $Y = f(t)$ would appropriately represent *percentage* change rate. Such an approach would be much better than calculating $\Delta N/N = (N_{t+1} - N_t)/N_t$ values (see Section 3.5.3 on ratios). Where observations at only two times are available (see Section 4.2) change rate calculated as Δ log N = log N_1 − log N_0 would be appropriate.

De Marche (1976) provides an example, from a study of species diversity in stream benthos, of the use of equation parameters in subsequent analysis. The intercept and slope from the linear regression of number of species (S) on substrate mean particle size in phi units (ϕ) are plotted over time and interpreted ecologically. Since phi units are a logarithmic transform of particle size in millimetres, the intercept has the useful interpretation: number of species in zero phi (1 mm) average particle size substrate. The slope may be interpreted as the degree of dependence of species diversity on substrate particle size.

Test statistics, such as F- and t-values or the probability levels associated with them, are not often used as variables, although there is no reason why they cannot be.

The strength of a bivariate relationship as measured by the Pearson correlation coefficient r can be used as a variable in subsequent statistical analysis (see Section 3.9 regarding the appropriate transformation). The probability clustering procedure of Goodall (1966b, c) uses probabilities of H_0: "no association between two variables" to calculate similarity coefficients. An example of informal use of F-values as measures of the strength of contribution of variables to groups derived from a cluster analysis is in Section 4.4

3.6 SELECTION OF VARIABLES

Choices must always be made about which variables will be measured and used for statistical analysis models in environmental studies. When biological variables are species, some subset of species in the community must be chosen, and the choice should always have a logical basis. Spight (1976), for example, argues for environmental studies based on key species, instead of surveys where some arbitrary number of species is considered. One approach is to consider the species variables as a sample of all possible species variables in the population—that is, in the natural community being sampled. This is analogous to the problem of sampling the phenome when choosing morphometric characters in numerical taxonomy (Sokal and Sneath 1963, Thorpe 1976), in that one hopes the variables chosen are a sufficiently large random sample for the statistical analysis results to be similar to what would have been obtained if all possible variables had been used. Unfortunately, it is probably even more difficult to randomly sample species in a community than to sample taxonomic variables in a phenome (Section 2.3.6). The question, "How do you choose a subset of species variables with most information about the entire community?" has two parts: (a) which species have the most information, and (b) how many of the most informative ones contain enough, say some specified percentage, of the total information? A second approach to the choice of variables is to ask what variables are of most interest in their own right (commercially important fish species, say) or carry the most information (per unit effort or cost perhaps) about the hypotheses. The general subject of variable selection is included in the review paper by Crovello (1970).

A few species may carry most of the information about the structure of a community in space and time. For example, Austin and Greig-Smith (1968) find that an efficient description of a rain forest can be obtained with an ordination based on less than 25 percent of the total flora. In multivariate analyses the smaller the number of biotic variables and of environmental variables the better, consistent with adequate description of both impact effects and natural background variation. There are five reasons why large numbers of variables in a multivariate analysis are not recommended. First, more variables make it easier to obtain a significant result, hence more difficult to interpret that result. Second, the sensitivity of multivariate tests to violation of assumptions tends to increase with the number of variables (Olson 1976). Third, more variables mean more measurements, more sorting, counting, and identification work, and higher computer cost. Fourth, the additional information obtained from additional species variables diminishes rapidly (Kaesler et al 1974), assuming that the most informative species are selected first. Fifth, more variables increase the chance that intractable matrices may cause the analysis to break down (see Section 3.5.3).

If one wishes to use many species as biotic variables, it is best to apply principal components (or principal coordinates) analysis to the log-transformed species abundance data, or to binary data, as illustrated in Section

3.4.3. The first $k < p$ principal components, sufficient to account for some specified percentage of the information in all p variables, can then be used as the biological variables for subsequent statistical analysis. Caution should be exercised when discarding small principal components, however. The variation associated with a principal component is a function of both the number of variables involved and the amount of variation in those variables. Also, a large amount of the variation may be irrelevant to the purposes of the study. Therefore, a small principal component could represent a single species variable that is relatively invariant and uncorrelated with other species, but is a faithful predictor (or indicator) of the impact. It is always a wise precaution to examine scatter plots of scores for biological principal components, even very small ones, against the environmental variables before discarding those PCs. This is easily done with such computer statistical packages as SAS (Service 1972, Barr et al 1976), SPSS (Nie et al 1976), and APL (1969). Orloci (1975c) discusses the differences among PCA, FA, and univariate methods in dealing with variation common to a number of species (the covariances among them) as opposed to the variation specific to each. Williams and Lambert (1959, 1960) describe and give examples of an "association analysis" procedure for presence-absence species data which is analogous to PCA and can be applied to fairly large data sets using only a small calculator (preferably with some programming capability). It is a particularly useful procedure, easy to understand and explain, for choosing the best indicator species. Wermuth et al (1976) describe computer programs for either quantitative or binary multivariate data that identify correlated variable subsets and provide a condensed description of the data.

To reduce variables to the number of the original variables containing the most information, the procedure for ranking variables by a dispersion criterion described by Orloci (1973a) is appropriate. This procedure uses a partial correlation technique to rank the variables according to the information each contains about the entire variable set, and then uses multivariate analysis of covariance (see Cooley and Lohnes 1962, 1971, and Green 1974) to remove each variable in turn and determine the remaining information. A simple example follows, which is based on the $n = 36$ by $p = 3$ simulated biological data set in Table 4.1 of Section 4.1. The dispersion matrix (of deviation sums of squares and cross-products) is based on log-transformed data:

$$\begin{bmatrix} 6.9941 & -1.6389 & 1.1126 \\ -1.6389 & 19.3372 & 2.9945 \\ 1.1126 & 2.9945 & 8.2062 \end{bmatrix}.$$

The dispersion criterion S_1 for the most informative variable is

$$S_1 = \max \left[\frac{6.9941^2}{6.9941} + \frac{-1.6389^2}{6.9941} + \frac{1,1126^2}{6.9941}, \frac{-1.6389^2}{19.3372} \right.$$
$$\left. + \frac{19.3372^2}{19.3372} + \frac{2.9945^2}{19.3372}, \frac{1.1126^2}{8.2062} + \frac{2.9945^2}{8.2062} + \frac{8.2062^2}{8.2062} \right]$$

$$= \max [7.5551, 19.9398, 9.4498]$$

$$= 19.9398,$$

which indicates that the second variable is most informative and 19.9398 is the measure of the information it contains. The residual matrix, where information about the second variable has been removed by multivariate analysis of covariance, is

$$\begin{bmatrix} 6.9941 & -1.6389 & 1.1126 \\ -1.6389 & 19.3372 & 2.9945 \\ 1.1126 & 2.9945 & 8.2062 \end{bmatrix} - \begin{bmatrix} -1.6389 \\ 19.3372 \\ 2.9945 \end{bmatrix} \begin{bmatrix} -1.6389 \\ 19.3372 \\ 2.9945 \end{bmatrix}^{TR} [19.3372]^{-1}$$

$$= \begin{bmatrix} 6.9941 & -1.6389 & 1.1126 \\ -1.6389 & 19.3372 & 2.9945 \\ 1.1126 & 2.9945 & 8.2062 \end{bmatrix} - \begin{bmatrix} .1389 & -1.6389 & -.2538 \\ -1.6389 & 19.3372 & 2.9945 \\ -.2538 & 2.9945 & .4637 \end{bmatrix}$$

$$= \begin{bmatrix} 6.8552 & 0 & 1.3664 \\ 0 & 0 & 0 \\ 1.3664 & 0 & 7.7425 \end{bmatrix}$$

Note that all matrix elements having to do with the second variable have gone to zero, and the other matrix elements have lost any covariation with the second variable that they contained. The dispersion criterion S_2 for the second most informative variable is

$$S_2 = \max \left[\frac{6.8552^2}{6.8552} + \frac{0^2}{6.8552} + \frac{1.3664^2}{6.8552}, 0, \frac{1.3664^2}{7.77425} + \frac{0^2}{7.7425} + \frac{7.7425^2}{7.7425} \right]$$

$$= \max [7.1276, 0, 7.9836]$$

$$= 7.9836,$$

which indicates that the third variable is the next most informative and 7.9836 is the measure of information in it, *given that* any information common to both it and the second variable has already been removed. Finally, the residual matrix where information about both the second and third variables has been removed is

$$
\begin{bmatrix} 6.8552 & 0 & 1.3664 \\ 0 & 0 & 0 \\ 1.3664 & 0 & 7.7425 \end{bmatrix} - \begin{bmatrix} 1.3664 \\ 0 \\ 7.7425 \end{bmatrix} \begin{bmatrix} 1.3664 \\ 0 \\ 7.7425 \end{bmatrix}^{TR} [7.7425]^{-1}
$$

$$
= \begin{bmatrix} 6.6141 & 0 & 0 \\ 0 & 0 & 0 \\ 0 & 0 & 0 \end{bmatrix}
$$

and

$$
\begin{aligned}
S_3 &= \max\,[6.6141, 0, 0] \\
&= 6.6141,
\end{aligned}
$$

which indicates that the measure of information contained uniquely (= not common to either of the other variables) in the first and least informative variable is 6.6141. Note that $\Sigma_1^p\, S_i = 34.5375$ is equal to the sum of the diagonal elements in the original dispersion matrix, indicating that the total variation has been correctly partitioned. The plots of percentage $S_{i=1,\,p}$ against the p variates in rank order of their residual information (Figure 3.13) show at a glance the percentage of information that would be retained by keeping the $m < p$ most informative variables and discarding the rest. This three-variable example is a trivial one, but it should be obvious that the method can be extended to any number of variables, and the format of Figure 3.13 can be used for deciding how many and which species variables to include—by including either those necessary to account for some specified percentage of the information in all variables or those preceding (= with more information than) the region where the $\%S_i$ curve starts flattening off. An example taken from Orloci and Mukkattu (1973) is shown in Figure 3.14. Confidence limits are calculated under H_0: species selected at random. Orloci (1976) presents a similar method based on an information criterion that can handle mixed data or data of any type. Orloci's (1975a) book provides computer programs in the BASIC language for both methods. Although it is not strictly possible to test for significance of the information in the least informative $p - m$ variables, some approximate tests appropriate to multivariate covariance analysis are given by Rao (1966).

   If the desired approach is to choose the variables that carry the most information about the hypotheses then one may proceed in several ways. The system and the variables to be measured in it can be chosen based on *a priori* knowledge or by applying some kind of variable selection procedure to the hypothesis-testing statistical model, using preliminary data.

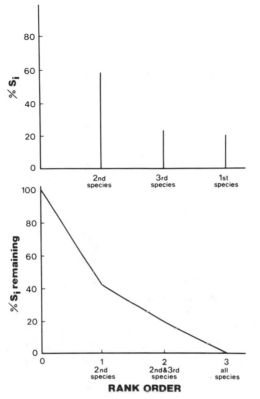

FIGURE 3.13 Percentages of the dispersion criterion $S_{i=1,p}$ as a function of the $p = 3$ variates in rank order of their residual information. The ordinate %S_i remaining, is the percent of the information (in all species) that is left out if only the species shown on the abscissa are used.

What criteria should be applied to choice of a system and variables in it for use in applied environmental studies? I suggest the following: (1) spatial and temporal stability in biotic and environmental variables that would be used to describe or predict impact effects, (2) feasibility of sampling with precision and at reasonable cost, (3) relevance to the impact effects and a sensitivity of response to them, and (4) some economic or aesthetic value, if possible. In marine and freshwater environments the benthic community is commonly chosen for these reasons (Keup et al 1966, Cairns and Dickson 1971). Keup et al discuss the importance that benthic studies have assumed in legal cases concerning environmental impact. Katz (1972) provides a good example, involving mercury and pike, of why highly mobile and difficult to sample nekton are usually a poor

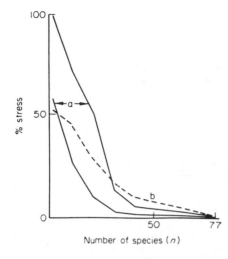

FIGURE 3.14 The solid lines (a) represent the .96 confidence limits for the "stress" (analogous to the lower part of Figure 3.13) resulting from the use of a given number of the highest ranked of a total of 77 species. The dashed line (b) is the lower .96 confidence limit for species taken in a random sequence which for these data suggests that, with up to 25 or 30 species, ranking species by the information they contain is no better than choosing the species at random. Reproduced with permission of the British Ecological Society from Figure 2 of Orloci and Mukkattu (1973).

choice even though they may represent most of the economic and aesthetic value in the biological community. Again using aquatic environments as an example, the species variables can be chosen based on the same criteria. Mollusk species, for example, are sedentary, long-lived, and often sensitive to pollution. Also, they take up important pollutants, are easily marked, are often possible to age, and leave their shells as evidence of location, and often time and age, of death (see Section 3.7). As a result of past environmental studies, summaries of indicator taxa have accumulated in the literature, for example, Keup et al (1966), Cairns and Dickson (1971), and Hart and Fuller (1974) in fresh water. Spence and Hynes (1971) group species by their sensitivity to impoundment effects, and Kelso et al (1977) by sensitivity to a pulp mill effluent.

Many variable selection procedures for statistical analysis models have been described and some may be valuable for the purpose of selecting variable subsets, based on preliminary data and for use in the main study. However, they are often misleading and often misrepresented. It should be emphasized that variable subset selection procedures rarely have a valid interpretation in terms of significance of the variables selected. This is true for ranking or selecting variables by their contribution in any statistical model, be it multiple regression analysis, discriminant analysis, or any other, and it is as true for interpreting the results of the main study as it is for selecting variables to use in the main study. Valid tests of significance of individual variables in predictive models assume that those variables are uncorrelated (hence the name independent variables), which is an often ignored assumption. If the variables are correlated, as they almost always are in an environmental study, not only the judgment of significance of individual variables, but also the evaluation of rank order of importance, are meaningless.

Visualize the following situation. An impact criterion species variable is predicted by four environmental variables, which are total dissolved oxygen, algal biomass, total phosphorus, and temperature. Assume that two different things are going on in the impact of an effluent on the community: nutrient enrichment and temperature effects. Assume further that the nutrient enrichment has twice the effect on the criterion species variable as does temperature. If there had been two independent predictor variables, say total phosphorus and temperature, the standardized coefficient for the former would have been twice that of the latter. When three highly correlated variables share the description of nutrient enrichment, each has a coefficient two-thirds as large as that for temperature. A significance testing procedure that assumes independence may consider only the temperature variable to be significant, and a deletion procedure may delete all but the temperature variable. The false conclusion that temperature is the most significant predictor, or even that it is the only worthwhile predictor, could be reached. This invalidity of any stepwise variable selection procedure applied to correlated variables is mentioned by Hope (1968) and Eisenbeis et al (1973) in their reviews of various procedures. Cassie (1972) describes an informal *ad hoc* procedure and makes the same point. Many stepwise variable selection procedures are available for both regression analysis and discriminant analysis, in SAS (Service 1972, Barr et al 1976), SPSS (Nie et al 1975), and BMD (Dixon 1973), and the fact that there are so many implies that different procedures give different variable importance rankings because there is no correct ranking with correlated variables. If the wrong statistical model is used, the variable selection procedure is even less valid. For example, Watt (1968) suggests the use of multiple regression as a general preliminary screening procedure for variable selection in environmental studies, and the appropriateness of the multiple regression model in most cases is doubtful (see Section 4.4).

One possible remedy is to first carry out a PCA as described above and then use the PC scores as observations on new uncorrelated variables in a subsequent statistical analysis. In that analysis significance testing of the PC variables would be valid, and variable-selection methods would be unambiguous. It should be noted that Orloci's variable ranking and selection procedures as described above would also be ambiguous if applied to correlated data for the purpose of providing a meaningful ranking of predictor variables according to their power in predicting a criterion variable. A covariance analysis is, in fact, one of the procedures used for significance testing and stepwise deletion of variables in predictive models (e.g., Horton et al 1968). A simple nonparametric discriminant analysis procedure described by Kendall (1966) yields a ranking of variables by their discriminatory power, but unambiguous interpretation again requires uncorrelated variables. Saila et al (1976) presents methods for variable selection in an environmental study where the choice is based on information obtained per unit cost. Riechert (1976) shows how samples or sites can be ranked in the same way as for variables.

What taxonomic level should be used for biological taxon variables? Identification of species is usually assumed to be the desirable though not always attainable goal. The major argument for this would be that the use of higher taxa containing more than one species that occur in the study area could result in an environmentally heterogenous population masquerading under the label of a single variable (see Section 3.5.3). Against this can be put two arguments. One is that if it is a choice—as it often is—between a reliable identification at the generic level and guesswork at the species level, then the generic level is preferable. A supposedly valid indicator species can turn out to be a mixture of sibling species in any case (e.g., Grassle and Grassle 1976). Another argument is that when a genus or family is a sensitive indicator of pollution but the species level taxonomy is poorly worked out or difficult for anyone but the specialist, using the higher level taxon is better than not using one at all. Keup et al (1966) emphasize that coarse taxonomic categories are often satisfactory. Vascotto (1976) compares multivariate analyses of lake benthic assemblages based on specific and generic levels, and his results suggest that the generic level results are not inferior. In fact the genera appear to reflect more the large-scale environmental patterns and the species the microenvironmental patchiness, suggesting that the generic level may be more appropriate for some environmental studies. Use of species level identifications for some groups may be like reading a map with a microscope. Pielou (1969) shows how the information content attributable to different hierarchical taxonomic levels in a data set can be determined, and such evaluation of data from preliminary sampling could be used to decide whether the work involved in finer taxonomic level identification is justified by the gain in information.

3.7 BIOLOGICAL VARIABLES IN SPECIAL AREAS

The presence or abundance of kinds of organisms are the most commonly used biological variables in environmental studies. Many others are also appropriate, however, or even more appropriate because of greater sensitivity to pollutants at sublethal levels, especially in laboratory monitoring situations. Certain kinds of environments, organisms, and pollution types call for particular variables and estimation procedures. Each of these topics has a voluminous literature, and an exhaustive review is not intended here. It is hoped that the reader will be provoked into considering alternative biological variables using various estimation methodologies, and can enter the literature for discussions or examples of their use.

Particular environments, types of impact, and groups of organisms require specialized sampling and estimation procedures. Southwood's (1966) book covers methods for sampling a variety of aquatic and terrestrial environments, and British Columbia Lands, Forests, and Water Resources (1974) covers such topics as sampling terrestrial vegetation, toxicant-specific systems, indicator species, aquatic sampling in various communities, and monitoring of dissolved gases. An improved sampling device and technique for stream benthos is described by Carle (1976). Holme and McIntyre (1971) present methods for the study of marine benthos. Coutant (1971) reviews methods for the study of thermal pollution.

Estimation of density, mortality and other population parameters for highly mobile organisms such as fish and birds may require mark-recapture methods. Ecological methods textbooks by Southwood (1966), Watt (1968), and Poole (1974) introduce mark-recapture procedures, with examples. Ricker (1958), Beverton and Holt (1957), Cormack (1968), Seber (1973) and Brownie et al (1978) are basic references. Unusual marking techniques applicable to particular organisms are described by Edmonson (1944), Swan (1961), and Rounsefell (1963). Dunn and Gipson (1977) provide a recent example of marking for radio telemetry. The following mark-recapture references and the organization of their presentation are after A. N. Arnason (personal communication).

1. Sampling at two times only:
   a. Estimation of size only, in one homogeneous population—the Peterson estimate. Use the basic references given above.
   b. Two noninterchanging populations (e.g., two sexes or species)—survey removal method. See Chapman and Murphy (1965).
   c. Several populations with migration among them but no mortality or recruitment—stratified Peterson. See Chapman and Junge (1956), Darroch (1961), and Arnason (1973).
2. Sampling at three or more times:
   a. Animals uniquely marked or remarked each time they are resampled; births and/or deaths may occur; the population is open—The Jolly-Seber model. See Jolly (1965), Seber (1965), Cormack (1972), and Arnason and Baniuk (1977).

b. Animals marked first time only. This is relatively inefficient and should be avoided, if possible. See Bailey (1951).

c. Population closed. Can mark first time only (Paloheimo 1963) or mark all unmarked individuals at each sampling (Otis et al in press).

3. Sampling at three or more times with information about catch effort at each time. General references are Ricker (1958) and Beverton and Holt (1957). The best estimates are for the models of Darroch (1958) and Paulik (1963).

Some extensions and tests for the Jolly-Seber model are provided by Carothers (1971), White (1971), and Robson et al (1972).

Organism size, shape, and growth rate can serve as biological variables that are sensitive to environmental factors and are easily measured. Sometimes they can be determined directly from long-lived organisms (e.g., tree rings—see Haugen 1967) or from artifacts left by organisms that have died (e.g., bivalve shell rings—see Clark 1968, Rhoads and Panella 1970, Tevesz 1972, and Green 1973). Well-documented museum specimens could be used. Modification of organism size and shape by both natural and man-caused environmental factors is well-known. Davis and Hidu (1969) and Hickey and Anderson (1968) report the effects of pesticides on embryonic development and growth in marine bivalves and on shell thickness in raptorial bird species, respectively. Induced thermal effects on diapause and growth in stream benthic species are described by Lehmkuhl (1972), and modification of fish shape because of a heated discharge by Stauffer et al (1974). Whittle and Flood (1977) assess growth impairment of fish by a pulp mill effluent.

Estimation of size structure and growth rates is extensively treated by Ricker (1958) for fish populations, but the methods are often generally applicable. Ricker considers mark-recapture methods for estimation of growth, and examples for organisms other than fish are in Edmonson (1944) and Frank (1965).

Determination of growth rate parameters requires the assumption of some growth model. Very specialized models have been proposed for various organisms, but the precision of most data rarely justifies such sophistication. Just as the most useful spatial distribution models are the generally applicable ones (see Section 2.3.9), so are easily fitted curves with approximately the correct shape for the growth patterns of most organisms. Probably the most widely used model is the von Bertalanffy or monomolecular curve:

$$S_t = S_\infty (1 - e^{-Kt})$$

where S_t is size at time t (given that $S_0 = 0$ at $t = 0$) and KS_∞ is the instantaneous growth rate at $t = 0$. The plot of S_t against t is a curve whose slope is the growth rate, which decreases linearly as S_t increases and reaches zero at the asymptotic size S_∞. In differential form this growth model is

$$\frac{dS_t}{dt} = K(S_\infty - S_t),$$

indicating that the growth rate is proportional to the growth remaining until the asymptotic size is reached. See Ricker (1958) for discussion and examples. Fitting this model to observations on S_t and t is complicated by the fact that the linear form is

$$\log\left(1 - \frac{S_t}{S_\infty}\right) = -Kt$$

and the value of S_∞ is usually unknown and must be estimated from the data. One approach is to use trial values of S_∞ until the best linear regression is found. Another approach is to use successive S_t and t values to calculate ΔS_t and Δt and then calculate the linear regression

$$\frac{\Delta S_t}{\Delta t} \approx K(S_\infty - S_t) = KS_\infty - KS_t$$

as an approximation to the differential form of the model. See Leveque (1971) for an example. A better approach is to use iterative nonlinear regression procedures (see Glass 1967, Conway et al 1970, Gallucci and Hylleberg 1976) which are now widely available in computer packages such as BMD (Dixon 1973). When ΔS_t and Δt are directly observed and Δt can be held constant, for example by using annual growth rings or by determining size in marked animals recaptured after equal time intervals, use of the Walford plot technique (Ricker 1958) is appropriate. The model is converted to a finite difference equation

$$S_{t+1} = a + bS_t$$

where $a = S_\infty(1 - e^{-K})$ and $b = e^{-K}$. Linear regression methods can be used and estimates of K and S_∞ can be calculated from the slope b and intercept a. Note that a and b, as well as S_∞ and K, are biologically meaningful growth parameters that could be related to environmental factors (see Section 3.5.4). While S_∞ is the asymptotic size and KS_∞ the instantaneous growth rate (at $t = 0$), the intercept a is the estimated first year growth and the slope b is the fraction of the growth remaining after the first year. Growth parameters can be contrasted between locations or times for existing or past populations (using artifacts such as shells for the latter) by using analysis of covariance. Such an analysis applied to the data shown in Figure 3.15 indicates that the null hypothesis of similar growth patterns at two intertidal locations should be rejected ($p < 0.01$).

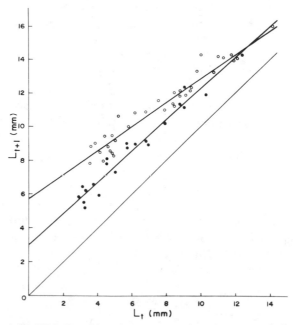

FIGURE 3.15 A Walford plot of lengths at consecutive winter growth rings for the clam *Macoma balthica* at high tide (open circles) and low tide (solid circles) levels in the Hudson Bay intertidal zone near Fort Churchill.

See Green (1973) for details. If the von Bertalanffy growth model is assumed and several other assumptions are made, both growth and mortality rates can be estimated by a graphical procedure from only observations on average size in a two-year class population at two times of the year (Green 1970a). See also Ebert (1973) for a generalization of this method with a computer program.

Levinton and Bambach (1970) suggest that an even simpler growth model is widely applicable, at least for bivalve mollusks:

$$S_t = k \log (t + 1).$$

For many other organisms as well this model may be adequate and the parameter k may serve as a measure of growth rate. See Section 3.5.4 regarding the use of equation parameters as biological variables. If one is in the opposite situation, with data indicating a growth form that only a more flexible model can fit adequately, then the flexible growth function of Richards (1959), which generalizes the von Bertalanffy and two other models, is suitable. Finally, for estimation of sizes at various ages from size distributions made up of several overlapping year classes, the procedure of Cassie (1954) is appropriate (see Section 3.5.3).

For multivariate size data (more than one morphological measure on each individual) principal components analysis is an appropriate first step. See Section 3.5.3 for discussion. The paper by Jolicoeur and Mosimann (1960) provides a particularly good example of the extraction of growth trend and shape components using PCA, and Atchley (1971) is an example of the use of factor analysis for the same purpose.

Relating size or shape measures, or growth rate parameters, to environmental predictor variables is no different than with any other kind of biological criterion variable. There may, in fact, be fewer difficulties because suitably transformed morphological variables are more likely to have an approximately linear relationship with transformed environmental variables than are species abundances. Linear statistical models such as multiple regression may be appropriate with the former, whereas they would not be with the latter (see Section 4.4). A multivariate linear model (canonical correlation analysis—see Cooley and Lohnes 1971, Marriott 1974, Harris 1975) is used by Green (1972) to relate a set of morphological variables to a set of environmental variables for explanatory prediction of shell shape variation in lake bivalve mollusks. The results are presented in Figure 3.16 where each closed curve encloses all samples from a lake or a region of a large lake. The ordinate represents a linear additive relationship between environmental variables correlated with water turbulence and morphological variables describing variation in overall size, and the abscissa represents a linear additive relationship between water chemistry variables and shell weight or thickness. The dashed curves are for two locations not included in the original analysis and represent successful predictions of environment from shell morphology. See Section 2.1.9 regarding the danger of possible confounding with morphological clines when using morphological variables in environmental studies.

Rao (1972) includes multivariate growth models in his review of recent trends in multivariate analysis.

Measures of productivity, energy flow, or biomass can be valuable biological criterion variables, although I agree with Hurlbert (1971) that arguments for such measures being fundamental in some unique way and therefore absolutely essential for ecological studies are baseless. The ecological methods texts by Southwood (1966) and Poole (1974) cover the estimation of production, biomass, and energy in different ecological systems. See also the International Biological Programme Handbook series. The recent book edited by Lieth and Whittaker (1975) has chapters on methods for productivity estimation in aquatic and terrestrial systems, measurement of caloric values, and assessment of regional primary pro-

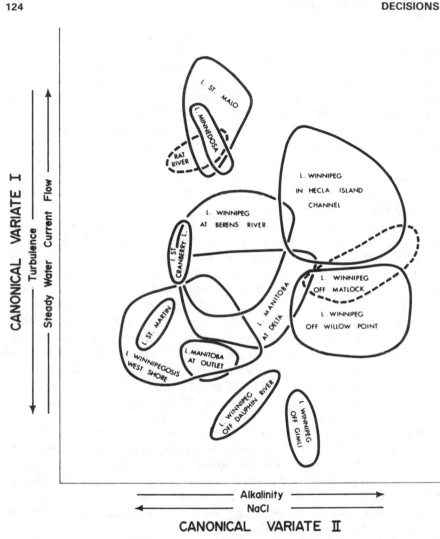

FIGURE 3.16 The distribution of lakes and lake areas by morphological type of *Lampsilis radiata* shell. The axes represent linear additive functions of morphological variables, derived from a canonical correlation analysis. Reproduced by permission of the Minister of Supply and Services Canada from Figure 2 of Green (1972).

duction. Winberg (1971) describes methods for estimation of production of aquatic animals, and Ricker (1958) covers estimation of secondary production related to fisheries. Holme and McIntyre (1971) present methods applicable to marine benthos. Combined laboratory and field studies often are necessary to estimate production in natural populations (e.g., Green and Hobson 1970). Some examples from applied environmental studies are Kamp-Nielsen (1971), Moore and Love (1977), and Kennedy and Mihursky (1972) who describe effects of mercury on *Chlorella*, pulp mill effluent on periphyton and phytoplankton, and temperature on bivalve mollusks, respectively. A rapid approximate method for estimation of annual production and turnover rate, which is particularly applicable to running water communities and is probably often adequate for comparative purposes, is described by Hamilton (1969) who reviews earlier methods of the same type. An example of the application of Hamilton's method is in Green (1971b).

Systems for continuous monitoring of respiration to evaluate sublethal effects of a pollutant on aquatic organisms are described by Livingston (1968, 1970), Hamwi and Haskin (1969), Hicks and De Witt (1970), and de Wilde (1973). The use of such systems for biological monitoring to detect impact (Section 4.3) is described by Cairns et al (1974) and by Westlake and van der Schalie (1977).

Behavior of organisms can be utilized in several ways for environmental studies. Passive sampling devices such as artificial substrates in aquatic environments (e.g., Crossman and Cairns 1974) and pitfall traps in terrestrial environments (e.g., Banerjee 1969) have advantages and disadvantages compared with direct sampling methods. Artificial substrates represent almost the only way natural substrate variation can be controlled in sampling benthic communities, which are often the best choice of system in aquatic studies (Section 3.6). Also, for most organisms it is the reproductive propagules or the juveniles that have the greatest dispersal and recolonization capabilities, and therefore passive sampling methods tend to sample the potential next generation of the species in the community rather than older individuals that became established when conditions were better, but cannot reproduce under present environmental conditions. On the negative side, aritificial substrates may not be typical of the natural environmental of the area (Section 2.3.6), and the distance from the artificial substrate to the source of the colonizing propagules is generally unknown. For example, samplers containing rocks can be colonized by stonefly larvae even when suspended over a muddy substrate.

Behavioral measures can be used as biological criterion variables that are even more sensitive than physiological measures, because organisms tend to first respond behaviorally to unfavorable conditions and then physiologically only if the behavioral response fails (Slobodkin 1968). Flannagan (1973b) found that large numbers of benthic species entered the drift after a stream drainage was sprayed with fenitrothion, although the standing crop did not change presumably because of rapid replacement from upstream. Kelso (1977) found that although fish tended to avoid a pulp mill effluent the numbers of fish in the general area were increased by the effluent, probably because of increased macroinvertebrate productivity. It should be remembered that behavior is also sensitive to environmental factors unrelated to impact (Saila et al 1972, Cowell and Carew 1976) and therefore must be used with caution as a criterion variable.

A third use of behavior in environmental studies is for assessment of sublethal effects of toxicants in the laboratory, and for predictive biological monitoring to detect an impact as described above for respiration rate. References other than those given above are Cairns and Dickson (1971), Cairns et al (1974), Livingston et al (1974), and Cooley (1977). Also see papers in the symposium volume edited by Cairns et al (1977).

The concentration of the pollutant itself in organisms can be used as a criterion variable, in some cases long after the impact occurred. The use of artifacts such as shells for this latter purpose was mentioned above in relation to size or growth rate as a criterion variable and will be considered further in Section 4.5. Although the organism may be of value in its own right, as with mercury in pike (Katz 1972), usually this approach depends on finding particular combinations of compounds and taxonomic groups known to selectively take them up (e.g., mollusks and heavy metals, tunicates and vanadium, certain aquatic insects and pesticides). For heavy metals in fish and shellfish see Merlini (1967), Miller (1972), Evans et al (1972), Cunningham and Tripp (1973), and Phillips (1976). Uptake and concentration of pesticides are described by Bedford et al (1968), Foehrenbach (1971), and Koblinski and Livingston (1975), of pulp mill effluent compounds by Kaiser (1977), and of petroleum hydrocarbons by Lee et al (1972).

Three cautionary remarks should be made. First, concentrations of some elements and compounds can be influenced by a variety of natural environmental factors, including climatic factors (Lee and Wilson 1969, Sturesson and Reyment 1971). Second, mobile organisms (such as fish) containing substances with long half-lives (such as mercury) may have been exposed to the substance at some unknown time before and distance away from where they are collected. Third, if concentration in an organism is taken to reflect concentration in the environment, it must be assumed that an equilibrium concentration in the organism has been reached. Laboratory studies to determine uptake and turnover rates for that compound by that organism are therefore advisable (e.g., Keckes et al 1969, Smith et al 1975).

3.8 ESTIMATION OF NECESSARY SAMPLE SIZE, NUMBER, AND ALLOCATION

Factors influencing choice of sample number and size were considered in Section 2.3.8, and examples of sample number estimation for two different designs were worked. Sample allocation in various designs was considered in Section 2.3.7. Some additional topics and methods related to sample size, number, and allocation are presented in this section.

Kuno (1969) describes a sequential sampling procedure (Section 2.3.9) which, instead of leading to decisions about whether means exceed given levels, is designed so that "sampling is terminated when a defined level of precision is reached" with precision measured by the ratio of the standard error to the mean (equivalent to estimating the mean within plus or minus some specified percentage of the mean). Using Kuno's notation, let

$$d = \text{standard error of the sample mean,}$$

$$D = d/m = \text{ratio of standard error to mean,}$$

$$T_n = \text{expected cumulative total for sample } n,$$

$$d = \sqrt{\frac{1}{n} f(m)}.$$

Therefore,

$$D = \sqrt{\frac{n}{T_n^2} f\left(\frac{T_n}{n}\right)}$$

where the variance $s^2 = f(m) = f(T_n/n)$, which is some variance-mean relationship, such as Taylor's power law $s^2 = am^b$ (see Section 2.3.9). Kuno proceeds with a variance-mean relationship related to the negative binomial distribution, which leads to decision lines (as in Figure 2.12) that are curves for all distributions except the Poisson. Green (1970b) proceeds with Taylor's power law, so that

$$D^2 = an^{1-b}T_n^{b-2}$$

and

$$\log T_n = \left[\frac{\log (D^2/a)}{b-2}\right] + \frac{b-1}{b-2} \log n,$$

unless $b = 2$ (implying the logarithmic series distribution) in which case

$$\log n = \log a - 2 \log D.$$

On a sequential sampling graph of T_n against n, with both ordinate and abcissa on log scales, the decision line is a straight line. The slope is zero if $b = 1$ (Poisson distribution), implying that sampling should continue until some fixed cumulative total number of individuals T_n is sampled, regardless of their density. For $1 < b < 2$ the line has a negative slope and for the special case of $b = 2$ the line is vertical, the latter implying that some fixed number of samples is collected regardless of the density.

An example is based on 35 sets of three replicate 43 cm² samples of small clams from a Queensland intertidal zone. The fitting of Taylor's power law was carried out as in Section 2.3.9. The regression of log s^2 on log m is highly significant (the H_0 of $b = 0$ is rejected at $p < 0.01$) and estimates of b and a in log $s^2 = \log a + b \log m$ are 0.90 and 2.80, respectively. A fixed value of D is chosen as $D_0 = 0.10$ (a standard error of 10 percent of the mean, implying .95 confidence limits of approximately plus or minus 20 percent of the mean) and the resulting decision line is shown in Figure 3.17. Three sets of replicate samples, each from a different tidal level with a different density, that had not been used in the calculation of the decision line are then employed in a sequential manner until the "stop line" is reached, and the three paths are shown with estimates $\hat{m}$ and $\hat{D}$ for each.

The fact that the logarithmic transformation usually suffices for species abundance data (see Section 2.3.9) implies that Taylor's power law with $b = 2$ is an adequate model most of the time. If this is so, some powerful generalizations about necessary sample number to achieve a desired preci-

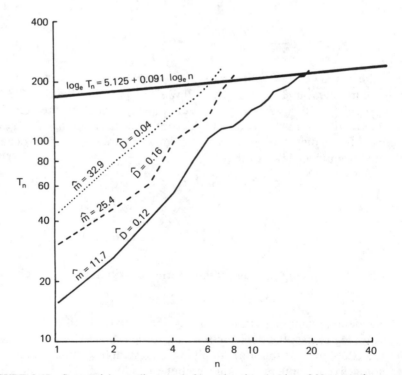

FIGURE 3.17 Sequential sampling graph for estimating density of *Notospisula parva* at Crib Island, Queensland with .95 confidence limits of ±20 percent of the mean. Both T_n (cumulative total for sample n) and n are on logarithmic scales. Reproduced with permission from Figure 1 of Green (1970).

sion can be made. These generalizations are stated here, and then some values from the literature are presented as a test of their validity. If

$$D_0^2 = an^{1-b}T_n^{b-2} \quad \text{and} \quad b \approx 2, \quad \text{then } n \approx aD_0^{-2},$$

$$a \approx 1, \quad \text{then } n \approx D_0^{-2}.$$

This states a very simple rule of thumb: for a wide range of field data the number of samples required to achieve a desired precision D_0 is independent of the mean density, and is approximately equal to the inverse of the square of the desired precision. If, for example, the mean density must be estimated with precision such that .95 confidence limits are plus or minus 20 percent of the mean, then $D_0 \approx 0.10$ and $n \approx 100$. If plus or minus 40 percent suffices, then $n \approx 25$ (see Figure 3.18). If $b \approx 2$ but $a \neq 1$, then the rule applies with the words "multiplied by a" added.

In his 1961 paper Taylor presents a and b values and the range of densities for 24 sets of field data taken from the literature, including shellfish, insects, fish, and other organisms. These data probably represent an attempt to cover a wide range of distribution types and probably contain more variable a and b values than would usually be encountered. Since

$$D_0^2 = an^{1-b}T_n^{b-2} \quad \text{and} \quad T_n = nm,$$

we have

$$D_0 = \sqrt{a/n}\, m^{\frac{1}{2}(b-2)}$$

$$\log D_0 = \log \sqrt{a/n} + \frac{1}{2}(b-2)\log m.$$

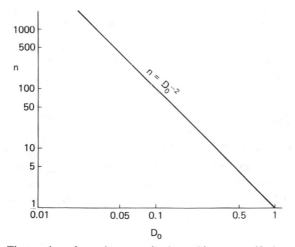

FIGURE 3.18 The number of samples n required to achieve a specified precision D_0 (ratio of standard error to mean) if $b = 2$ and $a = 1$ in Taylor's power law $S^2 = am^b$.

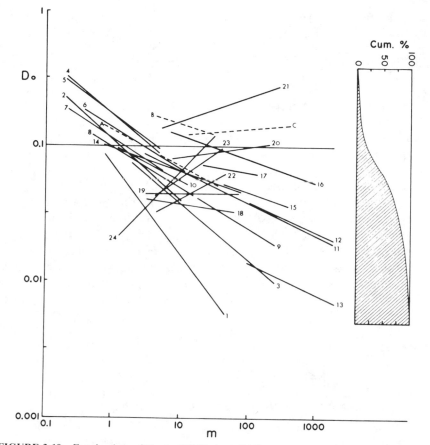

FIGURE 3.19 For the data of Taylor (1961) the solid lines connect the estimated precision (D_0) values that *would* have been obtained for the highest and lowest density samples in each data set if the "$b = 2$ and $a = 1$" rule of thumb had been used. The dashed lines are for three data sets which permit tests against the null hypotheses $b = 2$ and $a = 1$ (A: Queensland intertidal clams in 43 cm² samples; B: Pacific salmon in spawning streams; C: North Atlantic whiting in standard trawl tows). Both axes are on logarithmic scales.

For each data set I calculated the precision D_0 that would have been obtained for the highest and lowest density samples if the rule of thumb $n = D_0^{-2} = (0.1)^{-2} = 100$, assuming $b = 2$ and $a = 1$, had been followed. The solid straight lines in Figure 3.19 connect these pairs of D_0 values and cover the range of densities for each data set. The numbers correspond to Taylor's (1961) Table 1. The horizontal line at $D_0 = 0.1$ is the line that *would* be obtained for a data set where $b = 2$ and $a = 1$, because

$$\log D_0 = \log \sqrt{1/100} + \frac{1}{2}(2 - 2)\log m,$$

and therefore

$$D_0 = 0.1.$$

On the right in Figure 3.19 is a plot of the cumulative percentage of line lengths, which suggests that the rule of thumb based on $b = 2$ and $a = 1$ tends to be conservative, erring mostly in the direction of yielding $D_0 < 0.1$.

The values of a and b are *estimates* from sample means and variances for the data sets, and the departures of the lines as drawn, from the $D_0 = 0.1$ horizontal line, may not all be real. The dashed lines in Figure 3.19 are for three data sets which permit tests of significance of the null hypotheses $b = 2$ and $a = 1$. The a values are not significantly different from $a = 1$ in any set, and only the b value in set A significantly differs from $b = 2$. Therefore, only the set A line is significantly different from $D_0 = 0.1$, and the difference is such that the rule of thumb would tend to yield $D_0 < 0.1$. In summary, then, $n \approx D_0^{-2}$ appears to be an adequate approximate rule of thumb for achieving a desired precision D_0 for estimation of the mean density of biological organisms.

Necessary sample number for different specific purposes must be estimated by different methods. If the data are to be screened for outliers by methods such as those described in Section 3.9, a sample number of at least 30, and preferably 60, is required (Daniel and Wood 1971). Necessary sample number for correct rankings (with specified error probabilities) was discussed in Section 3.4.2. Livingston et al (1976) use simulation procedures for estimation of the number of subsamples required to obtain 90 percent of the total number of species in the primary sample, and also of the number of subsamples required to estimate mean biomass with a specified precision.

It was emphasized in Section 2.3.8 that, all other things being equal, it is better to sample the same total area or volume (the same total number of organisms on the average) by taking many small samples rather than few large ones. This is illustrated in the simulated sampling experiment shown in Figure 3.20 and Table 3.3. Square A contains a completely random distribution of 250 points, each point determined by coordinates drawn from a random numbers table. Square B contains an aggregated distribution of points, generated as follows. The first 50 points were generated at random. The other points were generated as follows. The position defined by a randomly drawn pair of coordinates was determined. If it would fall more than a specified distance (the length of a side of the smaller sample unit square at the top of the figure) from an existing point, it was not entered. If it would fall less than half that distance from an existing point, it was entered. If it would fall more than half the distance, but less than the full distance, from an existing point, it had a probability of 0.5

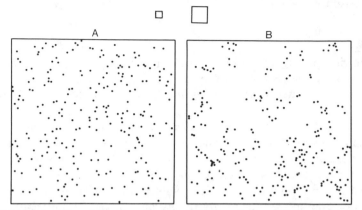

FIGURE 3.20   A random (A) and an aggregated (B) distribution of 250 points. The squares at the top are the two sample unit sizes used for random sampling of these two distributions.

of being entered. This process continued until 200 additional points had been entered.

Each distribution was sampled randomly using the sample unit sizes shown at the top of the figure. For the smaller sample unit size two sets of 100 samples from each distribution were obtained, and for the larger size (four times the area of the smaller one) two sets of 25 samples from each distribution. The results are shown in Table 3.3. For the random distribution (A) it makes no difference which sample unit size is used. The variances and means are all approximately equal to each other, as is expected for a Poisson distribution, and the ratios of the standard errors to the means are approximately the same. For the aggregated distribution (B) all variances are significantly greater than the means. Both the variance-to-mean ratios and the standard error-to-mean ratios are much larger for the larger sample unit sizes than for the smaller ones.

Dennison and Hay (1967) suggest that in benthic studies "most of the procedures currently in use produce at best semi-quantitative data", and they therefore use binomial sampling theory to estimate the number of individuals n which must be sampled so that a species present in proportion p in the community will have a probability of being missed no larger than some specified value Y. The result is the graphical estimation format shown in Figure 3.21. If the mean density is m, the appropriate sample unit size is n/m. It should be emphasized that the sampling must be strictly random, which in many communities may not be possible, for the binomial sampling theory to apply. Goodall (1961, 1973) states a rule of thumb for how large a sample unit size must be when there is spatial pattern, if the results of the statistical analysis are to be independent of the sample size.

Table 3.3 Population parameters and sample statistics for the distributions shown in Figure 3.20

Distribution	Random (A)				Aggregated (B)			
Sample unit size	1	4	1	2	1	4	1	2
μ	0.625	2.5	0.625	2.88	1.80	2.5	0.61	6.8
n	100	25	100			25	100	
μn	62.5	62.5	62.5			62.5	62.5	
Replicate sets	1	1	1	2	1	1	1	2
$\bar{x}$	0.55	2.80	2.88		2	1.80	0.80	6.8
s^2	0.49	2.42	2.69		4.25	1.25	44.3	
$s_{\bar{x}}$	0.070	0.311	0.328		0.412	0.112	1.33	
$s^2/\bar{x}$	$0.90ns$	$0.86ns$	$0.94ns$	1.32^*	2.36^{**}	1.56^{**}	6.52^{**}	
$100 s_{\bar{x}}/\bar{x}$	12.8	11.1	11.4	14.7	22.9	14.0	19.6	

$^*p < 0.05.$
$^{**}p < 0.01.$

133

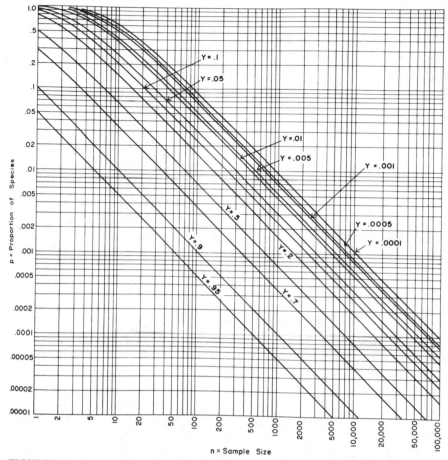

FIGURE 3.21 Number of randomly selected individuals n needed to detect a species whose abundance is a proportion p of the multispecies population, with probability of failure Y. Reproduced by permission of the Society of Economic Paleontologists and Minerologists from Figure 1 of Dennison and Hay (1967).

This minimal area should be "that of a square with the side equal to the distance at which the variance between samples ceases to be a function of their spatial separation." Alternatively, the sample unit size might be *decreased* until it is much *smaller* than the scale of spatial pattern, and the same rule of thumb could be applied.

Allocation of samples spatially was considered in Section 2.3.7, where stratified and nested designs were briefly discussed. Elliott (1977) provides an excellent concise treatment of the subject. The spatial allocation of the preliminary sampling should be over a large enough area that the distance of possible effects from the source can be estimated. Wind, water currents, or transportation by organisms can carry impact effects great distances in some cases. For example, Cleveland (1976) presents the results of a study showing how ozone concentrations caused by the interaction between New York City emissions and sunlight are highest over southwestern Connecticut, but elevated levels are carried by the prevailing winds as far away as northeastern Massachusetts. The spread of pesticide residues all over the world from their original locations of application, often through food chains and the movement of the organisms involved, is another extreme case (Wurster and Wingate 1968). Controls would be virtually nonexistent, in that many pesticide residues are now detectable in any food organism from anywhere on earth. In toxicological studies the same problem arises, and preliminary runs with levels set at broad intervals over a wide range should be carried out. When the approximate toxicant levels that bracket the range from "a bit more than zero" to "a bit less than 100 percent" mortality have been established, the levels can be evenly spaced over that range and thereby provide a very efficient design. As pointed out in Section 4.2, toxicant levels that produce 0 or 100 percent mortality do not contribute to estimation of LD_{50}'s.

Allocation of samples in time must be chosen with regard to the type of impact involved and in particular the response pattern of the biological community to that type of impact. There are many response patterns, and temporal allocation differs from spatial allocation in that it is rarely possible for preliminary sampling to provide the necessary information—except in the sense that response patterns for particular types of impact may be known from previous studies. Cairns et al (1971), for example, present four case histories of the recovery of damaged streams, discuss temporal patterns of response and recovery, and consider the factors on which speed of recovery depends. When the impact has not yet occurred and its location and time of occurrence are known (main sequences 1 and 2), the temporal dimension is in two parts: preimpact and postimpact. The preimpact part is analogous to a baseline study (main sequence 3), and both the duration (if there is a choice in the matter) and the intensity of preimpact sampling will largely be determined by the temporal patterns of fluctuation in the natural environment (e.g., seasonal, lunar, diurnal); these must be incorporated into the baseline model against which hypothesized impact effects would be tested. Livingston (1977), for example, discusses the importance of seasonal and other sources of variation in the community structure even of marine benthos, and the consequences for impact studies in which such variation is ignored. The so-called impact study where preimpact sampling is nothing more than a quick-and-dirty one-season survey is very common, even in cases where no prior knowledge of the community exists and the extent of year-to-year variation is unknown. Even in supposedly stable communities long-term studies show that natural year-to-year variation is often greater than the anticipated impact effects for which the initial one-shot survey was supposedly the temporal control. From studies of marine macrobenthos in subtropical Moreton Bay, Queensland, Stephenson et al (1976) conclude that " 'baseline' studies for environmental impact statements . . . should involve extensive chronological replication."

Postimpact sampling should begin as soon after the impact as possible to minimize interim nonimpact changes and should continue until impact effects are fully realized and possibly through the recovery period as well. Very slow response times, or response curves with complex patterns caused by long-lasting residues, call for a design with a long series of postimpact sampling times. See Section 4 for further discussion of both spatial and temporal allocation of samples in relation to particular main sequences and the analysis methods appropriate to them. A good general reference on some aspects of sampling in time is Munn (1970).

3.9   DATA SCREENING BEFORE ANALYSIS

Techniques for screening data prior to statistical analysis can be applied to data from preliminary sampling, to the main data set, or both. If the statistical analyses are for testing hypotheses, examination of the data for gross violations of the assumptions for those analyses is a wise precaution, and may be of value in choosing appropriate data transformations (see Section 2.3.9). If the statistical analyses are for exploratory descriptive purposes, such as determining whether the species composition of a community is spatially heterogeneous, or helping to choose variables for a subsequent analysis (Section 3.6), questions about the structure of the data set are in order: is there structure? Is it discontinuous, continuous, or both? How many dimensions (how many different things going on) would suffice for a description of the structure? Gower (1967a) emphasizes the value of visual presentation, throughout an analysis as well as for presenting results, in revealing features that might otherwise escape notice.

Serious violations of assumptions that may be detected at this stage are correlated errors (= residuals) and heterogeneity of error variation. If correlated residuals are detected in preliminary sampling data, a change of sampling design to ensure randomness of sampling is indicated—little can be done to correct the problem in the collected data (Section 2.3.3). Heterogeneity of variances may be caused by aberrant values resulting from gross errors or from extremely atypical samples that appear in visual displays as outliers, or it may be caused by dependence of variation on mean values. The latter suggests data transformation and the former suggests either removal of the outlier(s) from some or all analyses or the use of nonparametric (possibly rank) procedures. Testing for and describing variance-mean relationships and choosing transformations were considered in section 2.3.9. The emphasis here is on detection and treatment of outliers, with reference to the other topics where appropriate.

In discussing consequences of violations of ANOVA assumptions Cochran (1947) indicates that the effect of gross errors is serious, but he also emphasizes that habitual screening for and rejection of outliers leads to underestimation of errors. After all, normally distributed errors are sometimes far from the mean and Cochran argues that an outlier should be rejected only if (1) there is an *a priori* explanation of the aberrance (e.g., that samples from one location—which was missed in the preliminary sampling—differ greatly because of an unusual microenvironment, or that the sampler did not work properly), or (2) the value is impossible (e.g., it is a negative density, or a larger number of animals than the sampling device can hold). Problems caused by outliers in ordination procedures are discussed by A. J. Anderson (1971).

Scatter plots of residuals are invaluable for rapid visual inspection of the error structure of a data set. The SAS statistical package (Service 1972, Barr et al 1976) is particularly useful. It allows for the residuals about the predicted values in any statistical model (ANOVA, regression, covariance, etc.) to be kept as a new data set, which can then be displayed in scatter plots as residuals for different variables against each other or as residuals against expected values for each variable. Daniel and Wood (1971) describe and illustrate the use of residual plots for checking normality, spotting outliers, and verifying the adequacy of regression models. Computer programs are described. An example of evaluation of multivariate statistical procedures from residual plots was given in Section 3.4.2 and Figure 3.5.

Plots of residuals against each other and against expected values for the species abundance data (Table 4.1) used in the example of Section 4.1 are shown in Figure 3.22. The expected values are the means for each of the four areas. Because they are simulated data, they are known to be lognormally and independently distributed, without gross errors (see Section 4.1). Part A of the figure suggests that the variation is heterogeneous and dependent on the mean (logarithmic transformation will correct this), and part B shows no evidence of correlated errors. Nowhere are there outliers that would indicate gross errors; the number of samples is only marginally adequate for outlier detection, however (Section 3.8).

When there are many variables it is useful to first carry out a principal

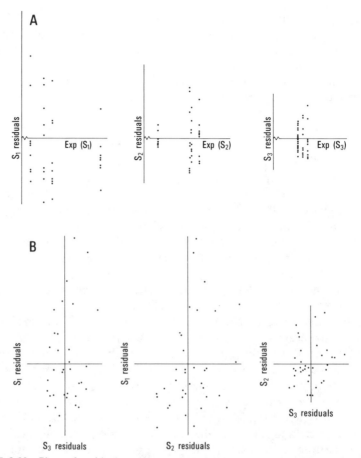

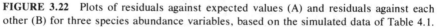

FIGURE 3.22 Plots of residuals against expected values (A) and residuals against each other (B) for three species abundance variables, based on the simulated data of Table 4.1.

components or principal coordinates analysis (Sections 3.4.3, 3.5.3, 3.6) on the data to compress the greatest possible variation into the first few principal components and then scatter plot the residuals on these PC variables. An example is in Cassie (1972). Projection of samples onto scatter plots on PC axes is easily done in the NT-SYS statistical package (Rohlf et al 1974).

If an apparent outlier is detected and the criteria of Cochran mentioned previously are satisfied, what then? Cochran (1947) and Dixon (1950) describe tests for univariate outliers and criteria for their removal from the data set. For the multivariate case Orloci (1972) and Bradfield and Orloci (1975) discuss the measurement of distance between a sample and a group of samples to which it may or may not belong. Testing for multivariate outliers is discussed by Wilks (1963), Gnanadesikan and Kettenring (1972), and Rohlf (1975).

Outliers can represent important information, not just gross errors that obscure the information sought. For example, suppose that baseline levels of a predictor variable have been determined and monitoring of the level continues at regular intervals. If an observed level is an outlier, with less than some probability α of occurring by chance alone given the baseline distribution of levels, an environmental impact of some kind may be indicated. See Section 4.3 for further discussion and examples.

If the goal is description of structure in a data set, the order of questions should be the following. (1) Can H_0: "no structure" be rejected? (2) To what extent is the structure continuous or discontinuous? (3) How many different (i.e., statistically independent) things can you justifiably assume to be going on in that data set? When answering these questions Occam's Razor (Section 2.1.2) should be kept in mind. Your data set *may* contain much informative structure. However, descriptive multivariate statistical methods can describe supposed structure that is nothing more than random variation, and it is easy to convince yourself that things you want to see in the data *are* in the data. A factor analysis will extract factors and a cluster analysis clusters from any data set, even random normal deviates.

Because one would expect that approximately 5 percent of correlation coefficients in random data would be "significant" at $\alpha = 0.05$, it is a wise precaution to start a search for structure with an overall test against H_0: "all correlations are zero." The sphericity test of Bartlett (1954) is appropriate. If there are n samples and p variables and R is the p-by-p correlation matrix, then

$$X^2 \left[\frac{p(p-1)}{2} \, df \right] = - \left[n - 1 - \frac{2p+5}{6} \right] \ln |R|.$$

It is a powerful test, and following Occam's Razor one should not be too impressed by a significant result. A nonsignificant result, however, indicates that one has no business searching for structure in that data set. For example, the test against H_0: "all correlations are zero" for the residuals shown in part B of Figure 3.22 is

$$X^2 \left[\frac{3(3-1)}{2} df \right] = - \left[36 - 1 - \frac{2(3)+5}{6} \right] \ln \begin{vmatrix} 1 & .24 & .33 \\ .24 & 1 & .07 \\ .33 & .07 & 1 \end{vmatrix},$$

$$X^2 \,(3 \, df) = - \left(35 - \frac{11}{6} \right) \ln 0.8396$$

$$= 5.7985ns.$$

Therefore, it can be assumed that the residuals are uncorrelated (which is true, given the properties of the simulated data).

Although a multivariate normal distribution is assumed, with large sample sizes the test should be adequate as an approximation even for non-normal data. This test extends to a test of significance of the $p - k$ principal components remaining after removal of the k largest PCs, and as such represents an approximate large-sample test of "how many different things are going on?" given that there is structure at all. Harris (1975) is a good reference for those tests, which in some cases can be carried out on either the correlation matrix or the covariance matrix. This is an important distinction when an ecologist is testing for structure in a p-by-p matrix where the data are log-transformed species abundances that are not otherwise standardized (see Sections 2.3.9 and 3.5.3). In a correlation matrix all diagonal elements are 1's and therefore significant structure implies among-sample variation in percentage species composition. Variation in abundance but with the same percentage species composition would not contribute to significance, although it would if the covariance matrix (which has species variances in the diagonal) were used. Which should be used is an ecological value judgment rather than a statistical question. My own feeling is that abundance variation can often be as informative about environmental impact as variation in percentage species composition. However, it is useful to know how to statistically separate the two. Factor analysis techniques (Cattell 1965a, b) also allow extraction of components of variation, with approximate tests of their significance (see Section 3.4.3).

The test against H_0: "no structure of any kind" should be accompanied by visual inspection of the data plotted on pairs of variables or on pairs of PCs as described previously. Answers to the question "how many things are going on?" could be posed in terms of clusters if there are apparent discontinuities in the data (e.g., Cassie 1972). Wolfe (1970) describes a nonhierarchical clustering procedure called NORMIX, which assumes that any structure is in the form of a mixture of overlapping multivariate normal distributions. Either common or differing covariance matrices (i.e., cluster sizes and shapes) can be assumed. The best fit to one, then two, and so on, multivariate normal clusters is found and the significance of H_0: "k clusters" versus H_A: "$> k$ clusters " is assessed at each step using a log-likelihood criterion. The assumption of normality can probably be approximated by transforming the observations on the original variables to principal components before entering the NORMIX program. Goodall (1966b, c) describes "probability clustering" procedures that allow approximate tests of the number of groups. These tests assume independence of the original variables, and therefore a PCA transformation is again a useful preliminary step.

Use of ordination and clustering methods in concert for assessment of whatever continuous or discontinuous structure may be in a set of data is discussed by Gower (1966a) and Goodall (1973b). This is a more sensible approach than setting ordination and clustering in opposition to each other, as some workers have done. The latter attitude is even less tenable since Lefkovitch (1976) has shown the close relationship between PCA and hierarchical cluster analysis (Section 3.4.3).

Interpretation of ordination axes as explanatory environmental factors is another matter, and should only be done with caution. The linear, or at least monotonic, species-environmental variable relationship assumed by the application of most ordination models to ecological data is often inappropriate (Sections 2.3.9, 3.4.2). See Section 4.4 for further discussion.

Some other approaches to assessment of structure and the minimum dimensionality required to describe that structure will be briefly mentioned. Borucki et al (1975) describe the use of cluster analysis, not for finding clusters but for detecting associations among sets of variables that may represent nonlinear as well as linear relationships. They recommend the use of multidimensional scaling (see Section 3.4.2) for description if such associations are found. Hills (1969) describes graphical techniques for examining large correlation matrices for structure. Correlation coefficients can be transformed to standard normal deviates as

$$z = \tfrac{1}{2}\log\left[(1 + r)/(1 - r)\right]$$

and then plotted on probability paper, to assist in detection of any $r \neq 0$. Detection of groups of interdependent variables from plots of principal coordinates analysis (Section 3.4.3) results is then described. Crovello (1970) argues that "for many descriptive purposes in ecology and systematics [the minimum spanning tree] is the best possible summary of relationships in a multidimensional character space." A minimum spanning tree is that network of straight lines connecting all sample points which has the minimum total line length. Rohlf (1970) describes its projection down to a three dimensional PC space, and computer programs are available in the NT-SYS package (Rohlf et al 1974).

3.10 COMPUTER PROGRAMS

This list is not intended to be exhaustive. However, procedures discussed in this book can be executed by one or more of these programs. No attempt has been made to cite references to these programs in other sections of the book because of the number of citations that would be necessary in some cases. Some programs and especially packages of programs are applicable to different analyses, so the arrangement here is by categories of accessibility: generally available packages, programs available from institutions, programs available from the author on request, listings in books or published papers, and reviews of available programs. Within these categories the programs are arranged so that similar types of analyses are grouped together.

1. Statistical packages and interactive systems:
   a. The Statistical Analysis System (SAS). Manuals are by Service (1972) and Barr et al (1976), and include lists of installations where SAS is available.There are flexible procedures for data input, transformations, multivariate and other statistics, graphics, time series analysis, clustering, and calling up procedures and files from outside SAS (e.g., BMD, SPSS). In my opinion, this is the best statistical package. Unfortunately, it is designed for IBM installations and is usually not available at others.
   b. The Statistical Package for the Social Sciences (SPSS). The manual is by Nie et al (1975). As with SAS, data files can be loaded, manipulated in various ways, and used as input to various procedures. A variety of statistical analyses and graphic procedures are included. Although recent versions are improved, the procedures (especially multivariate) are more limited than SAS and the control card format is more cumbersome. This is probably the most widely available and used package.
   c. Biomedical Computer Programs (BMD and BMDP) from the University of California, Berkeley. See Dixon (1973, 1975). These are widely available and include just about every commonly used analysis procedure in existence. However, they are really a collection of programs with differing and often difficult to use control card procedures, and it is best to use SAS or SPSS procedures if they exist for what you want to do. If they do not exist, try BMDP programs first, as they are easier to use than the old BMD programs.
   d. CLUSTAN is a package of clustering procedures developed by Wishart (1975). The manual (a computer printout) gives advice on procedure selection, which is fortunate because the forest of options is bewildering, various combinations of options are incompatible, and the error diagnostics are not good. This may be improved in a new version just made available.
   e. NT/SYS is a system of multivariate statistical programs developed by Rohlf et al (1974). Procedures for clustering, ordination (including PCA and nonmetric multidimensional scaling), matrix manipulation, and graphics are included.

f. APL/360 is an IBM interactive terminal language. The user's manual is APL (1969). It is easy to learn and use, and is particularly useful for teaching (Wilkinson and Huesman 1973) and applying multivariate statistics because matrices and vectors can be manipulated as if they were single numbers. Some versions of APL have limited work spaces and no interfacing with card input or output. Libraries of statistical routines, graphics routines, and so on, are available, but there are odd gaps (e.g., no good eigenvector routine).

g. BASIC is a simplified programming language analogous to FORTRAN, and is often used in interactive time-sharing systems. A reference is Boillot and Horn (1976). Matrix statements can be used effectively for multivariate analyses. The multivariate programs of Orloci (1975), for example, are all in BASIC.

2. Programs available from institutions or individuals who developed them:

a. The Ecological Analysis Package (EAP) can be ordered from Dr. R. W. Smith, 553 1/2 Avenue A, Redondo Beach, Calif. 90277. It is written in PL1 and includes programs for matrix manipulation, simulation, missing value estimation, graphic display, clustering, PCA, multidimensional scaling, several ordination procedures, discriminant analysis, regression analyses, canonical correlation, and diversity indices.

b. The Cornell Ecology Programs (Gauch 1976) are Fortran IV programs with extensive descriptive commentary, for data screening, analysis of sample and variable similarities, multivariate data simulation, and various ordination procedures.

c. A series of computer programs by Ellis (1968) for descriptively summarizing ecological field data.

d. Programs described by Daniel and Wood (1971) for data screening, variable and residual plots, and description of intervariable relationships.

e. Two scientific subroutine packages for matrix and vector manipulation are SSP (International Business Machines Corp. 1970) and EISPACK (Smith et al 1976). The latter is recommended for matrix roots and vectors solutions. Together they contain all the subroutines necessary for creation of virtually any multivariate analysis program.

f. GRAPHPAC graphic output subroutines written for the GE 635 computer (Rohlf 1969), for depicting spanning trees, scatter plots, perspective views of response surfaces, stereo pairs.

g. Programs for computation and plotting of response surfaces (O'Leary et al 1966).

h. PROMENADE, an interactive graphics pattern-recognition system for display in up to three dimensions on a CRT terminal (Hall et al 1968).

i. LOGIST and STLOG (Lee 1974), programs for linear logistic regression analysis after Cox (1970).

j. The Multivariate Statistical Analyzer (Jones 1964), a system of Fortran II programs that examines univariate distributions and bivariate relationships in multivariate data, and plots all significantly nonlinear bivariate relationships.

k. TMS, Fortran IV programs for baseline modeling and event detection in time-series analysis (Bower et al 1974).

l. POPAN-2, a data maintenance and analysis system for mark-recapture data (Arnason and Baniuk 1977). Emphasis is on Jolly-Seber open models.

m. A Fortran program for the Jolly-Seber multiple mark-recapture model (White 1971).

3. Programs available on request from the author (I have not verified current availability):

a. A Fortran listing of a technique for fitting nonlinear models to biological data (Glass 1967). See also Conway et al (1970).

b. A program for the method of Ebert (1973), modified from Green (1970a), for estimating growth and mortality rates in a mixed year class population given mean individual size at different times of the year.

c. A program for fitting the truncated lognormal distribution to species-abundance data by maximum likelihood (Slocomb et al 1977).

d. A program to partition data into qualitative and "additional due to quantitative" components (Noy-Meir 1971).

e. A program for ranking variables by an information criterion (Orloci 1976).

f. Programs for identifications of subsets of correlated variables in quantitative or binary data sets (Wermuth et al 1976).

g. A program for generating random normal multivariate data with specified means, variances, and correlations (Capra and Elster 1971).

4. Program listings in publications:

a. Orloci (1975a)—programs in the BASIC language for analysis of multispecies data, including similarity/dissimilarity coefficients, ranking of variables by information content, and various ordination procedures.

b. Anderberg (1973)—various programs for clustering.

c. Jardine and Sibson (1971)—programs for clustering, including an overlapping cluster model.

d. Cooley and Lohnes (1962, 1971)—programs for a wide variety of multivariate statistical analyses including principal components and factor analysis, multivariate analysis of variance and discriminant analysis (one-way and factorial designs), multivariate analysis of covariance, canonical correlation, and classification procedures (probability ellipses).

e. Davis (1973)—FORTRAN IV programs for simple univariate and bivariate statistics, matrix operations, linear interpolation (for contouring), graphic display, polynomial regression, time series analysis, response surface analysis, discriminant analysis, cluster analysis, PCA and factor analysis.

f. Lee (1971)—programs for principal components analysis, discriminant analysis, canonical correlation and canonical trend surface analysis. Some of these programs require debugging.

g. Carle (1976)—program for removal sampling estimate, designed for stream benthic studies.

h. Cairns and Dickson (1971)—program for calculation of information-based diversity indices.

i. Hocutt et al (1974)—program for calculation of diversity index $\bar{d}$.

5. Reviews of computer programs and packages:

a. Press (1972) reviews computer programs for multivariate models, including ANOVA, covariance, canonical correlation, discriminant analysis, clustering, matrix manipulation, latent structure analysis, principal components and factor analysis, and multidimensional scaling.

b. Enslein et al (1977) is an up-to-date guide to computer statistical methods, including chapters on simulation, variable selection in regression, stepwise regression and discrimination, factor analysis, multivariate ANOVA and covariance, cluster analysis, multidimensional scaling, and time series analysis.

In summary, the environmental biologist with the ability to use SAS or SPSS, CLUSTAN, and APL or BASIC, will find that most of his or her computing needs are met, though there may be need to turn elsewhere for certain clustering and ordination procedures and for some time series analysis. I find it useful to have also FORTRAN subroutine decks for each of the following: (1) calculation of a p-by-p covariance matrix from an n-by-p data set, with transformation and standardization options, (2) matrix multiplication, (3) matrix inversion, (4) roots and vectors of a symmetric and (5) a nonsymmetric matrix. Most multivariate analyses are possible with these hooked together in various sequences.

3.11   VISUAL DISPLAY OF RESULTS

Effective formats for presentation of results are important for communicating to lay audiences the implications of the study. It is sometimes necessary to do this in a courtroom where one must communicate to people who distrust statistics—often with good reason, unfortunately—but will believe their eyes. In addition to the need to get the results across, it should be remembered that clarity of thought (or its lack) tends to carry through all the stages of a study. Unclear presentation of results is likely to make the audience suspect that the conception and execution of the study were no better.

Some examples of effective graphical format were given in Section 2.1.8. Good general references on effective display of results are Lewis and Taylor (1967), Crovello (1970), Munn (1970), and Batschelet (1976). Bivariate scatter plots were discussed in Section 3.9 in relation to data screening. Computer systems and packages such as GRAPHPAC (Rohlf 1969), APL (1969), BMD and BMDP (Dixon 1973, 1975), SAS (Service 1972, Barr et al 1976), and SPSS (Nie et al 1975) allow this to be done easily with large data sets. In SAS different symbols can be used for different groups, and if quantiles of ranked data (upper third, middle third, and lower third, say) on any additional variable are created and defined as the groups, effective contour plots can be produced. Cleveland and Kleiner (1975) describe and give examples from a program "for enhancing scatter plots with moving statistics" which also divide the data into quantiles. References to the use of computer programs for contour plotting are O'Leary et al (1966), Gittins (1968), Siccama (1972), and Lieth and Whittaker (1975). With smaller data sets interpolation by hand is feasible (e.g., Fincham 1971). Contours drawn by eye around quantile values of canonical variate scores from multivariate analyses are shown in Figures 3.23 and 3.24, taken from Smith (1976). Contour plots are really two-dimensional representations of response surfaces, which can also be displayed as three-dimensional perspective views. Response surface methodology is reviewed in a recent paper by Mead and Pike (1975). See also Chapter 6 of

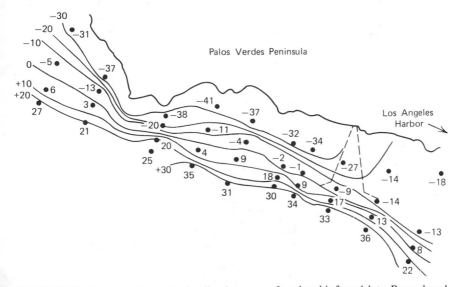

FIGURE 3.23 Contours for levels of ordination scores from benthic faunal data. Reproduced with permission from Figure 8-20 of Smith (1976).

Davis (1973). Rohlf (1969) describes a computer program for perspective representation of response surfaces. Results of toxicological studies (Section 4.2) can be displayed effectively in this way, either as contour plots (Figure 4.11), or as perspective views (Figure 4.13). Figure 3.25 is an example, taken from Davis (1973), of a technique for representing a four-dimensional trend surface in a three-dimensional perspective view.

Three-dimensional perspective views can be used effectively in many ways. The relation of a dendrogram to the spatial and temporal design of the study is shown in a perspective view in Figure 4.2. Allen (1971) presents the results of an ordination on algal species in a perspective view (Figure 3.26) with regions and environmental information superimposed. James (1971) uses a perspective view of pins at different heights to show similar data, and Carmichael and Sneath (1969) illustrate the use of ball-and-wire models. Figure 3.27 is an example of a stereo pair, taken from Fraser and Kovats (1966), which will appear genuinely three-dimensional if three-dimensional glasses are used. An example is also in Rohlf (1970). The GRAPHPAC (Rohlf 1969) computer subroutines can produce such stereo pairs. Higher-dimensional data can be displayed in two dimensions by using symbols that give information about the other dimensions (Gower 1967a, Whittington and Hughes 1972). The two-dimensional slice of a $p > 2$-dimensional space which contains the maximum possible information

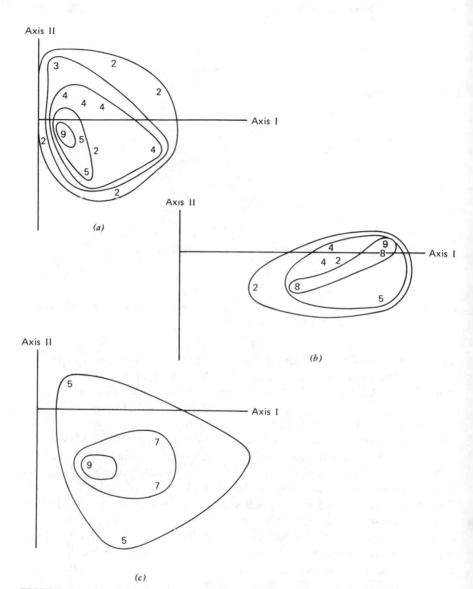

FIGURE 3.24 Contours for abundances of three species in the space defined by the first two discriminant functions of environmental variables. (*a*) *Lucinoma annulata;* (*b*) *Chloeia pinnata;* (*c*) *Lumbrineris Index*. Reproduced with permission from Fig. 8-18 of Smith (1976).

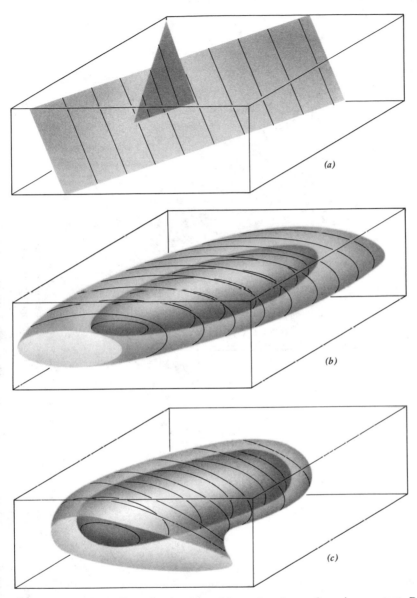

FIGURE 3.25 Four-dimensional polynomial trend surfaces of uranium content. Darker, inner envelopes have higher uranium content. The trend surfaces obtained from first, second, and third degree polynomials are shown in (a), (b) and (c) respectively. Reproduced by permission of John Wiley and Sons from Figure 6.21 of Davis (1973).

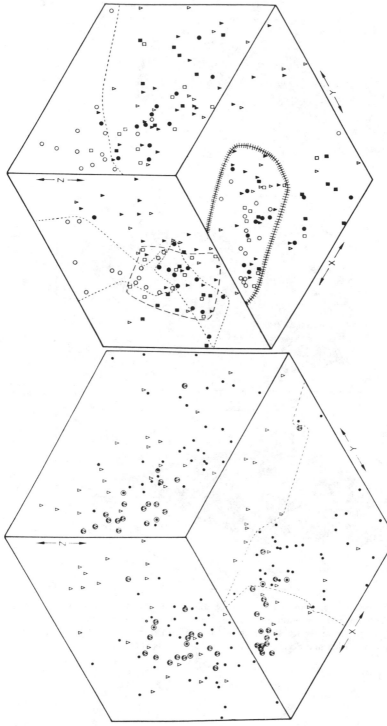

FIGURE 3.26 On the left are shown sites distributed in a three-dimensional ordination (PCA) space, for algal species on rock surfaces in Wales, with different symbols representing different moisture regimes. On the right are the same ordination results but with geographic regions indicated by symbols and boundaries on the *yz* facet. The line on the right *xy* facet is a subjective discontinuity used to define a second smaller set for further analysis. Reproduced by permission of the British Ecological

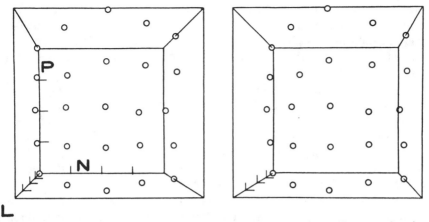

L

FIGURE 3.27 A response surface for length of tops of Douglas fir seedlings as a function of nitrogen and phosphorus levels, displayed as a stereo pair. Reproduced by permission of the Biometric Society from Figure 1 of Fraser and Kovats (1966).

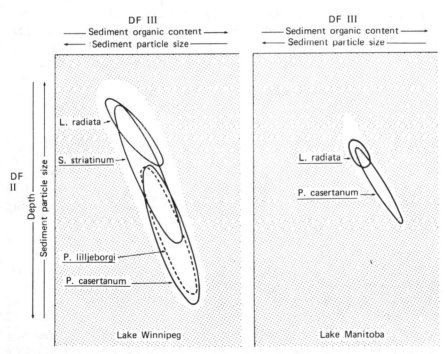

FIGURE 3.28 For two lakes the 0.5 probability ellipses for four mollusk species are shown. The axes are linear additive functions of environmental variables, derived from a discriminant analysis. The clear regions represent the values of discriminant scores for samples containing mollusk species. Reproduced with permission from Figure 5 of Green (1971).

about all *p* dimensions is of course that described by the first two (largest) principal components (see Section 3.4.3). The results of cluster analyses and ordinations by PCA, or other methods such as discriminant analysis, can easily be combined as in Figure 4.18. A procedure in the CLUSTAN computer package (Wishart 1975) permits groups derived from cluster analysis to be displayed in PC space, with or without convex hulls around the groups. Graphic output of publishable quality can be produced.

The bivariate equivalent of vertical bars used to represent confidence limits on univariate values (e.g., Figure 2.9) is probability ellipses. For example, the ellipses shown in Figure 3.28, taken from Green (1971), are 0.5 probability ellipses. For each group approximately half the samples are inside the ellipse and half outside it. Such probability ellipses can be calculated on pairs of the original variables or on pairs of derived variables such as principal components and discriminant functions (as in Figure 3.28). The calculation of the ellipses is straightforward and is described by Cooley and Lohnes (1971), Davis (1973), and Batschelet (1976). See also Section 4.3. Dunn and Gipson (1977) use probability ellipses to describe home range. Concentric convex hulls are used by Moss (1967) and Rohlf (1970) to describe closeness of relationship in cluster analysis results.

A variety of techniques for displaying the degree of relationship among a large number of samples (or of variables) are reviewed by Carmichael and Sneath (1969) and McIntosh (1973). One of the oldest and most effective formats is the trellis diagram, and several examples are given here. Figure 3.29, taken from Bloom et al (1972), shows among-time and among-location faunal similarity values in the lower left triangle and five similarity categories as types of shading in the upper right triangle. Figure 3.30, taken from Tietjen (1971), is similar but also includes environmental (depth and substrate) information along the margins. Figure 3.31 from Gage (1972) shows both faunal similarities (lower left triangle) and environmental similarities (upper right triangle) as dots of different sizes. The mirror image effect suggests a strong environmental-faunal relationship. Figure 3.32 from R. S. Anderson (1971) is an interesting trellis diagram format showing relationships among species and other relevant biological information. Other techniques for displaying relationships among samples or variables include network diagrams (Thorpe 1976), minimum spanning trees (Rohlf 1969, 1970, Clifford and Stephenson 1975), and dendrograms (or trees). Examples of the last are Figures 3.11, 4.2, and 4.17.

The mapping of biological and environmental variables onto spatial and temporal dimensions provides an effective display of pattern, for example related to impact. Howell and Shelton (1970) is a good example of mapping spatial patterns of fauna in relation to environmental impact (china clay waste). Margalef (1958b) uses mesh size in a network of lines to show among-location faunal similarities in an estuary. Figure 3.33, taken from

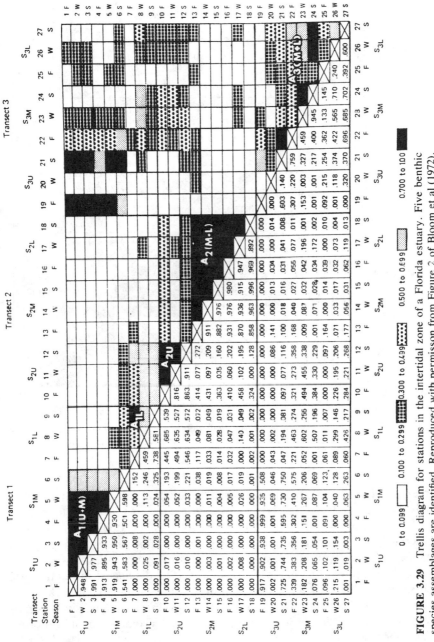

FIGURE 3.29 Trellis diagram for stations in the intertidal zone of a Florida estuary. Five benthic species assemblages are identified. Reproduced with permisson from Figure 2 of Bloom et al (1972).

153

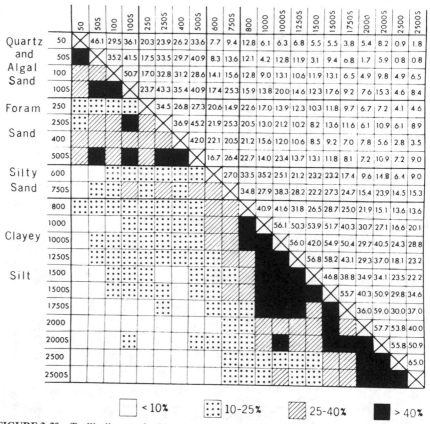

FIGURE 3.30 Trellis diagram for 21 stations, with depth in meters and substrate information given. Percentage faunal affinity for nematode species is shown. Reproduced with permission from Figure 3 of Tietjen (1971).

Green and Hobson (1970), shows faunal assemblage groups from a cluster analysis mapped onto an intertidal area with superimposed tide level contours. Mapping onto a temporal dimension can be done with symbols (Fincham 1971, Hunt and Jones 1972), and spatial patterns of temporal change are effectively displayed in a spatial-by-temporal plot of species-assemblages derived from a cluster analysis, as discussed in Section 4.2 and shown in Figure 4.8. Display techniques for variable values changing over time are discussed in Munn (1970). Periodic functions, such as polynomials and Fourier series, can be used to describe seasonal and other cyclic phenomena (see Batschelet 1976). They can also be used to describe spatial variation, as in Figure 3.34 taken from Curtis (1972).

As mentioned in Section 3.4.3, data consisting of percentages, fractions,

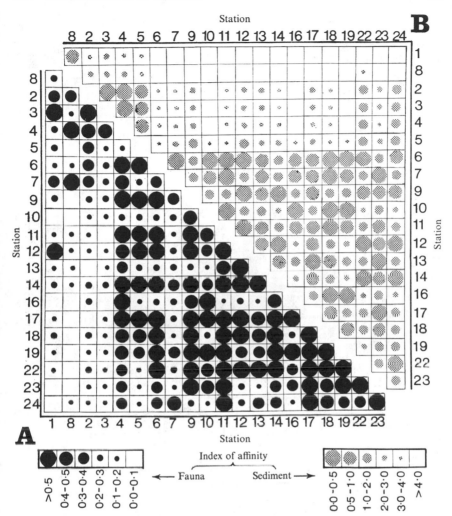

FIGURE 3.31 Trellis diagram for benthic faunal similarities among stations (A) in Loch Etive, and substrate composition similarities among the same stations (B). Reproduced by permission of Cambridge University Press from Figure 5 of Gage (1972).

or probabilities in three categories can be effectively displayed on triangular plots (e.g., Figure 3.9). See also Gower (1967a), Bryant and Sokal (1967), and Batschelet (1976). Ranks can be displayed using mapping techniques for projection of spherical surfaces onto planes (see Section 3.4.2).

This ends Section 3 on decisions. The sequences section (Section 4),

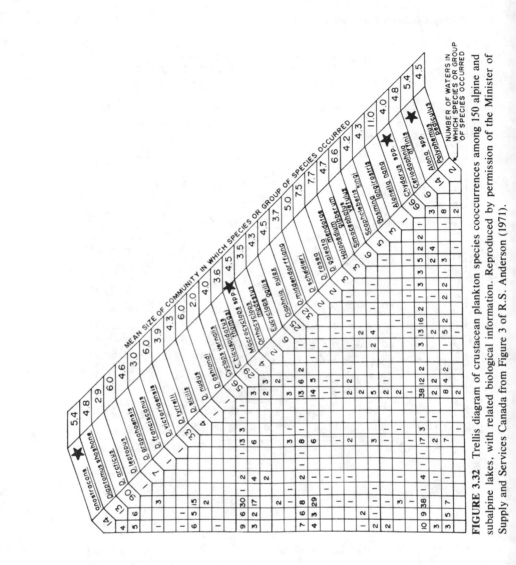

FIGURE 3.32 Trellis diagram of crustacean plankton species cooccurrences among 150 alpine and subalpine lakes, with related biological information. Reproduced by permission of the Minister of Supply and Services Canada from Figure 3 of R.S. Anderson (1971).

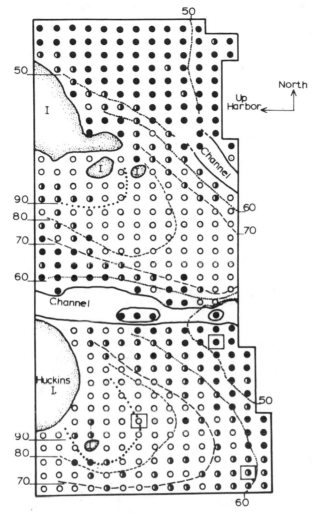

FIGURE 3.33 Three faunal assemblage groups derived from a cluster analysis of intertidal species presences and absences in samples on a 10 m interval grid. The three types of circles represent the three groups, and the contour lines represent centimeters above mean low water. Reproduced with permission from Figure 1 of Green and Hobson (1970).

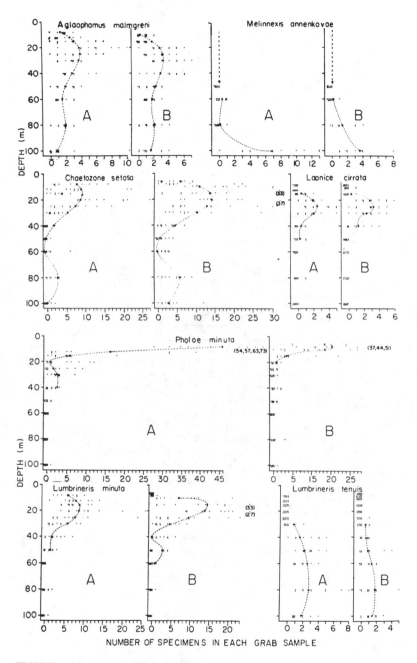

FIGURE 3.34 Polynomial curves fitted to depth distribution data for polychaete species in two fjords (A and B). Reproduced by permission of the Minister of Supply and Services Canada from Figure 1 of Curtis (1972).

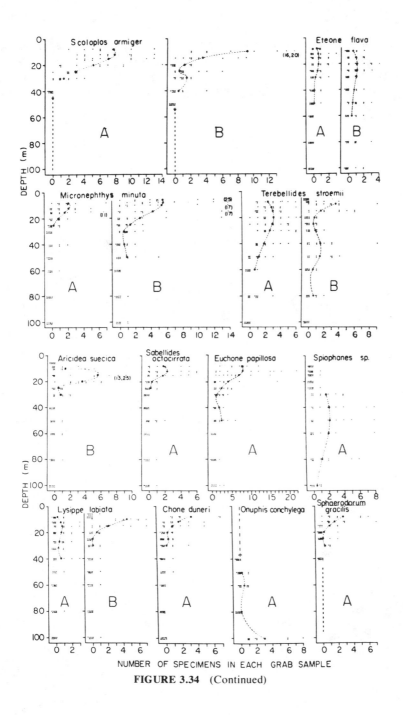

FIGURE 3.34 (Continued)

which follows, is organized by the five main sequences representing approaches to different objectives in environmental studies (Sections 3.1 to 3.3). Decisions that must be made within any of the main sequences (Sections 3.4 to 3.11) are discussed or illustrated by examples throughout Section 4.

Four

SEQUENCES

Optimal impact study design (Section 4.1) was discussed in Section 3.2 in the context of the spatial-by-temporal framework that defines an environmental study (Section 3.1). Also discussed were the prerequisites for such a design. In Section 3.3. other designs that form the suboptimal main sequences (Sections 4.2 to 4.5) were examined.

4.1 MAIN SEQUENCE 1: PERMITS AN OPTIMAL IMPACT STUDY DESIGN

Optimal sampling design and statistical analysis options are given in rows 3 and 4 and columns 3 to 6 of Figure 3.2. An impact study example for the optimal design is described in Section 2.2, and then it is used in Section 2.3 to illustrate the principles of sample design and statistical analysis. One biological criterion variable (abundance, rank abundance, or presence-absence for one species) was related to an impacted versus non-impacted environmental predictor variable implicit in the factorial AN-OVA design (see Sections 2.1.5 and 2.3.1). In an example later in this section the addition of two more species variables (as abundances) produces three biological criterion variables, which are related to the impact environmental variable implicit in the same design. The result is a *multivariate* factorial ANOVA. A measure of impact-related biological change is derived from this analysis, and then related to a measure of impact-related environmental change derived from a similar analysis on three explicit environmental predictor variables that are affected in various degrees by the impact. This example is also presented in Green (in press).

This sampling and analysis design (and that of the example in Section

2.3) is in row 4 and column 3 of Figure 3.2, representing one sampling time before and one after the impact and more than one location per each of two areas (control and impact). The statistical analysis model is that of the last row of Figure 3.3, with the right-hand column describing the univariate example of Section 2.3 and the left-hand column describing this multivariate example. Here we also illustrate spatial allocation in a nested design (Section 2.3.7), in contrast to the completely random design of Section 2.3. Other optimal designs, with more than one time of sampling before or after impact and differing numbers of areas or of locations per area (see Figure 3.2) are discussed in the course of the following presentation of this example. Use of the derived linear additive functions of the biological and environmental variables as measures (or indices) of impact (see Section 3.5.1) for subsequent biological monitoring to detect a future impact of the same type is illustrated in Section 4.3.

The optimal impact study design is often not possible because the environmental biologist is funded after-the-fact when someone complains about impact effects, or because no area suitable for a control area exists, or for some other reason. For such situations suboptimal designs are necessary (Section 3.3) and these are considered in the sections that follow (Sections 4.2 to 4.5).

In the two-way (areas-by-times) factorial multivariate ANOVA design the test against the null hypothesis of no change in the biotic community which is unique to the impact area is the test against the null hypothesis of no areas-by-times interaction for the biotic variables, as it was in the univariate case. The same null hypothesis may be tested for environmental change, using the environmental variables. In this analysis design one may allocate samples within the four areas-by-times combinations in either of two ways: (1) randomly or (2) randomly within each of a number of randomly allocated locations nested within the areas-by-times combinations (Section 2.3.7). The latter design allows control of among-location within-area variability caused by a heterogeneous environment, and is used in this example. Examples of the use of higher order multivariate ANOVA designs are provided by Mager (1974) and Harris (1975). Neither example is for an environmental problem, however.

It is also possible to convert the data to a form suitable for analysis in a one-way multivariate ANOVA design (as in the left column of the third row of Figure 3.3), if locations are randomly allocated in each area and then the same locations are sampled at each time as a basis for pairing. (This is in contrast to the factorial design where sample locations are randomly allocated in each of the four areas-by-times combinations.) For each location k and species j calculate the change in log-transformed species abundances N_{jk} as $\triangle_t \ln(N_{jk} + 1)$. The logarithmic transformation appropriately describes species abundance change as percentage change (see Section 3.5.4) and the addition of 1 allows the log-transformation of zero abundance values. Two groups (the impact and control areas) containing observations on species change variables at randomly allocated locations are thus produced. A one-way multivariate ANOVA will test the null hypothesis that the mean values of the $j = 1, p$ species change variables (for p species) do not differ between the two areas. This analysis design is analogous to that in a paired t-test, and similar to that used by Green (1977) and illustrated in Section 4.2, except that there is a control area as well as an impact area.

This design has several potential advantages over the factorial design, but also some disadvantages. The advantages are primarily those of the simpler one-way design over the factorial design in a multivariate ANOVA. Computer programs are more widely available and in the event of serious violation of the most critical assumption, that of homogeneity of within-area variance-covariance matrices, alternatives are available such as non-parametric procedures (Kendall 1966, Mantel 1970) and exact procedures for a valid test with heterogeneous matrices (Bhargava 1972). Also, it would be possible to regress the species change variables on environmental change variables, which is not possible in the 2×2 factorial design. However, there is the disadvantage that the random sample is equivalent to the random location, which is fixed for the two times. The influence of an odd location may be exaggerated because it will be sampled twice (i.e., at the two times). Sampling error will tend to be magnified by the conversion of the original log species abundances to differences, because the variance of a difference is equal to the sum of the variances of the original values. Finally, the factorial ANOVA design is appealing in that it directly and logically corresponds to the sampling design.

For those who wish to review multivariate ANOVA and related subjects such as canonical analysis and discriminant analysis, see Marriott (1974) and Harris (1975). Two earlier books by Cooley and Lohnes (1962, 1971) are useful references that also provide computer program listings. See also Section 4.4.

There is a varied methodology that falls into the general category of time-series analysis (column 6 in Figure 3.2). As the name implies, a series of observations in time is required, sufficient to establish the nature of such temporal pattern as may exist in the before-impact data. One may then test against the null hypothesis of no change in this temporal pattern after some specified time when impact begins. Optimal sampling design for time-series analysis (rows 3 and 4 of column 6 in Figure 3.2) of course requires that there be a control area as well, subject to parallel analysis and testing. See Section 4.2 for further consideration of time-series analysis. In a sense the ANOVA sampling design and analysis based on two times bypasses rather than describes any natural temporal variation by closely bracketing the impact event with sampling done just before and just after. For many environments with substantial seasonal variability and with impacts other than point-impacts in space and time, it is likely that multivariate time-series analysis methods will prove to be of great value. However such studies call for specialized temporal sampling designs with very large numbers of samples. This example emphasizes generality, with a sampling design and statistical analysis that can be widely applied and can in many situations satisfy the requirements of an optimal design with less sampling. The sampling design may be varied in several ways. More than two areas can be used, and the number of locations, the number of replicates per location, and the number of variables can be chosen to be appropriate for a particular study. This is discussed later.

As with any statistical analysis method, the validity of the multivariate ANOVA and tests of hypothesis based on it depend on certain assumptions. The assumptions are essentially the same as those for a univariate ANOVA (see Section 2.3.9), but in the multivariate case the number of ways in which the assumptions can be violated increase so rapidly with the number of variables p that a violation becomes highly probable. For example, with p variables (where $p = 1$ represents the univariate case) there will be p error distributions and therefore a much greater chance that at least one variable will not be normally distributed. However, non-normality is not a serious violation either for the multivariate ANOVA tests themselves (Harris 1975, Mardia 1971) or for methods that depend on canonical variates derived from them (Marriott 1974, Cooley and Lohnes 1962, Klecka 1975).

Heterogeneity of within-group variance-covariance matrices, the multivariate equivalent of heterogeneity of variance, is a more serious problem (Marriott 1974, Korin 1972). Heterogeneity may lead to an increase in the Type I error rate, and because there are $\frac{1}{2}p(p + 1)$ variances and covariances which could differ among groups the probability that at least one will do so rapidly approaches unity with increasing p (see Section 3.6). Multivariate tests for heterogeneity, analogous to Bartlett's test in the univariate case, are much more sensitive to the violation than are the multivariate ANOVA tests for differences in means which they justify (Marriott 1974, Mardia 1971). Therefore, they are not recommended.

Nonindependence of errors caused by correlated observations is always a serious violation in univariate or multivariate analyses (Sections 2.1.5, 2.3.3, 3.9). The variables on which the observations are made may of course be correlated, for example abundances of two species that have similar or different habitat preferences, but correlation caused by nonrandom allocation of replicate samples is a fundamental violation. The violation is easily avoided by making the necessary effort to assure that at some level the sampling is truly random. The design chosen here employs randomization at two levels: locations within areas and replicate samples within locations.

Additivity of effects is an assumption that is not likely to pose a problem. If it is violated, more serious violations will almost certainly occur. If those are corrected, any nonadditivity probably will be as well.

There are three reasons why violation of assumptions for a multivariate statistical analysis such as that recommended here should not frighten away potential users. First, regarding the most serious potential violation Marriott (1974) comments that "the tests break down only when variance heterogeneity is large and obvious." Also, it is tests based on a small sample number that are most sensitive (Section 2.3.8). Therefore use of a large sample number and then examination of the data for large and obvious violations are simple precautions. Second, heterogeneity and other violations are usually rooted in some dependence of variances and covariances on the means and the violations may be corrected by choosing appropriate transformations (see Section 2.3.9). Finally, the worst thing that could happen in case of a serious violation of the heterogeneity assumption is that significant among-group differences could be caused by differing variances and/or covariances rather than by different means. However, these are also group differences and are as likely to be interpretable and meaningful as are differences in means. In our example significant areas-by-times interaction for the biotic variables, caused by heterogeneity of variance-covariance matrices, would imply that within-location variation in abundances of or correlations among species had changed as a result of the impact. Certainly this could be interpreted as being an impact effect, just as change in species mean abundances can be.

Arguments for the use of simulated data to evaluate statistical methods were given in Section 2.1.7. Suppose that on a given date a sewage outfall will begin to release effluent at a point on the bank of a river (Section 2.2). An impact study is to be designed, using the benthic community (see Section 3.6) as the experimental system. As before, an area below the outfall serves as the impact area, while a comparable area above the outfall is the control area. The two areas are selected to be as internally homogeneous and as comparable to each other as possible, but assume that there is unavoidable environmental patchiness—say, the substrate varies among locations within areas from muddy to sandy. Therefore in this case we sample in a nested design by randomly allocating locations (three in this example) within each of the four areas-by-times combinations. At each of the 12 locations thus allocated we take three random samples, for a total of 2 areas $\times$ 2 times $\times$ 3 locations $\times$ 3 replicates = 36 samples. This sampling design is shown diagrammatically in Figure 2.6.

Three species $S_{j=1, 3}$ are of interest. Let the vector of their initial mean abundances per sample be

$$[N_1 \quad N_2 \quad N_3] = [100 \quad 50 \quad 25]$$

and let their within-location standard deviations (reflecting small-scale patchiness and sampling error) be 50 percent of those means. Let the within-location correlation matrix be

$$\begin{bmatrix} 1 & 0.6 & 0 \\ 0.6 & 1 & 0 \\ 0 & 0 & 1 \end{bmatrix}$$

implying that species S_1 and S_2 have some symbiotic relationship or a common microhabitat preference. Let the mean change due to impact be $[+50\% \ -75\% \ -30\%]$, with the result that in the impact area the vector of mean abundance per sample after impact becomes [150 12.5 17.5]. Of the three species, S_1 could be called a pollution indicator, while S_2 and S_3 are deleteriously affected in different degrees.

Let variation in substrate type be represented by a mud-sand gradient variable proportional to mean particle size in phi units (negative logarithm to the base 2 of particle size in millimeters). Let the 12 location values for this variable, hereafter referred to as the patchiness variable, be random normal deviates with mean zero and standard deviation 1. For each location let the species mean abundance vector be changed by $[-10\% +30\% -5\%]$ of the deviation from the mean value of zero of the patchiness variable. Of the three species, S_2 increases in abundance in the muddier (= finer = higher phi = higher value of patchiness variable) sediment, while S_1 and S_3 increase in differing degrees as the sediment becomes sandier (= coarser = lower phi = lower value of patchiness variable).

Let there be three environmental variables that are of interest, the level of each of which is measured for each of the 36 samples. Variable E_1 is percent organic content of sediment, with mean value of 10 percent where there is no impact effect. In the impact area, after impact, the mean value is increased to 40 percent. The within-location standard deviation is everywhere 10 percent, and there is a 0.6 correlation with the patchiness variable. Variable E_2 is mean particle size in phi units, with mean value of 2.50, a standard deviation of 0.5, and a correlation coefficient of 0.8 with the patchiness variable. Variable E_3 is oxygen concentration at the sediment surface in milligrams per litre, with mean value of 6, standard deviation of 1, and a correlation coefficient of -0.8 with percent organic content of sediment (E_1).

These are intended to be realistic data, with properties that would be characteristic of observations on such variables. For example, organic content of sediment would be expected to increase under the impact of organic pollution. As a consequence oxygen levels would decline. However, sediment organic content is also positively correlated with muddiness (i.e., with particle size in phi units) which varies naturally, and oxygen concentration will therefore show some correlation with natural variation in sediment type. We end up with a set of three biotic variables that interact with each other and with three environmental variables in various ways. As might be expected, the effects of the impact on the biota are both direct and indirect.

The data simulation procedure is described in some detail so that it can be repeated by anyone wishing to do so. All of the operations described were carried out on a Texas Instruments SR-52 programmable calculator. Where large or repeated simulation jobs are required, the operations could easily be done using a computer package such as SAS (Service 1972, Barr et al 1976).

The first step was to generate 36 observations on three species abundances, with specified means, standard deviations, and correlations. Independent random normal deviates ($\mu = 0$ and $\sigma = 1$) were converted to normally distributed variables with the specified means, standard deviations, and correlations, using the method of Capra and Elster (1971). The means were for logarithmically transformed values so that data with the skewed distributions and nonlinear relationships typical of species abundances could be generated. Standard deviations for the transformed variables were chosen appropriate to the desired standard deviations in the species abundances (see Section 2.3.9). Finally, the among-location variability due to substrate patchiness was introduced. To do this 12 additional random normal deviates were generated to represent the values of the patchiness variable at the 12 locations. Then each $\ln(N_j + 1)$ value was calculated as the prepatchiness value plus the product of the appropriate fraction for species j times the patchiness variable value for that location. These final $\ln(N_j + 1)$ values (36 observations on each of three species) were then back-transformed to give the simulated data, as in Table 4.1.

The observations on the three environmental variables were generated similarly, beginning with independent random normal deviates, using the method of Capra and Elster to introduce specific means, standard deviations, and correlations for appropriately transformed mean values, and then back-transforming these data. Variables E_2 and E_3 were calculated

Table 4.1 Simulated biotic and environmental impact study data. Biotic variables: abundances per sample for species $S_{j=1,\,3}$; environmental variables: E_1 = percent organic content of sediments, E_2 = mean particle size (phi units), and E_3 = oxygen concentration at the sediment surface (mg/l).

Sample No.	S_1	S_2	S_3	E_1	E_2	E_3
Impact area before impact						
A-1	73	14	23	3.4	1.35	8.5
A-2	130	36	14	1.2	1.74	9.0
A-3	70	14	24	27.3	1.95	3.5
B-1	114	39	17	11.5	2.68	6.2
B-2	158	91	36	22.9	1.81	4.3
B-3	152	68	13	0.3	2.37	9.6
C-1	42	14	20	5.3	1.73	6.2
C-2	103	87	27	1.6	2.17	7.5
C-3	63	26	33	26.3	1.74	3.8
Control area before impact						
D-1	58	39	18	11.5	2.43	5.8
D-2	59	73	21	1.9	2.94	6.6
D-3	73	33	16	15.4	2.73	5.9
E-1	85	50	51	1.3	1.78	7.3
E-2	83	33	31	4.0	2.76	7.8
E-3	166	28	30	8.7	1.95	6.1
F-1	139	53	63	0.5	1.79	8.3
F-2	48	23	16	14.1	2.14	5.7
F-3	84	61	38	0.4	2.46	9.8
Impact area after impact						
G-1	147	10	21	47.3	3.13	4.4
G-2	106	11	18	45.9	2.50	5.2
G-3	136	26	37	35.2	2.71	7.0
H-1	136	13	38	51.9	2.63	4.0
H-2	127	10	25	50.2	2.25	5.9
H-3	182	7	16	53.0	2.59	4.1
I-1	272	21	32	56.9	2.24	4.1
I-2	129	8	7	54.4	1.78	4.1

Table 4.1 *Continued*

Sample No.	S_1	S_2	S_3	E_1	E_2	E_3
I-3	149	10	13	33.4	2.07	6.3
Control area after impact						
J-1	68	64	19	5.9	3.23	6.1
J-2	52	57	14	49.0	3.20	3.8
J-3	69	88	15	18.9	3.32	5.1
K-1	163	53	19	17.2	2.15	6.8
K-2	79	34	38	2.0	2.06	9.1
K-3	86	29	48	1.6	2.06	6.1
L-1	125	45	14	3.4	1.99	7.5
L-2	123	29	32	8.8	1.75	5.6
L-3	214	57	50	0.1	2.25	10.7

for log-transformed means, but variable E_1 (percent organic content of sediment) was calculated for means based on a binomial distribution. Variables E_1 and E_3 were back-transformed but variable E_2 was not, because sediment particle size expressed in phi units is in logarithmic form. These simulated environmental data are also given in Table 4.1.

Let us now forget that these data have known properties and pretend that they are 36 observations on six variables from an impact study. Assume that the three species have been chose *a priori* by appropriate criteria applied to preliminary sampling. Perhaps they were the three species ranked highest in information content by one of Orloci's (1973a, 1976) procedures (see Section 3.6) or perhaps they had previously been described as good indicator species for the anticipated type of impact. The three environmental variables are assumed to have been chosen so that the environmental variable set will respond in an interpretable manner to the impact itself (e.g., percent organic content), to deterioration in any environmental variable likely to have a proximate effect on the biota (e.g., dissolved oxygen concentration), and to any natural spatial pattern in the environment upon which the impact effects will be superimposed (e.g., sediment mean particle size).

Because this example was designed so that it would be easy for the reader to repeat the calculations, the sample number (36) is unrealistically low. As a result some of the significance tests will not have sufficient power to reject null hypotheses that are known to be false in these simulated data. When these Type II errors occur the known properties of the data will override the significance tests.

All multivariate statistics calculations were carried out using the SAS package. Some matrix and vector output was modified so that the results could be presented in a different format, and these calculations were done on the SR-52 calculator. No computer programming was required. Programs for the factorial multivariate ANOVA used here can also be found in Cooley and Lohnes (1971), in BMD (Dixon 1973), and in the newest SPSS version just made available.

If the biotic and the environmental variables are appropriately transformed, any violations of assumptions are not likely to be serious enough to invalidate the conclusions. The robustness of the multivariate ANOVA has been discussed previously. With any statistical analysis it is a wise precaution to examine univariate histograms and bivariate scatter plots of the residuals (variation within locations in this case) both before and after applying transformations (Section 3.9). Nonparametric analogues to multivariate ANOVA and discriminant analysis do exist (see above) but they have not been developed for factorial designs. The observations on each variable could be replaced by their ranks if the sample number is large. Here the 36 simulated observations on six variables are transformed as we would transform real observations on such variables. Species abundances would be transformed as $\ln(N_j + 1)$, and concentrations (e.g., dissolved oxygen) would also be logarithmically transformed. Sediment particle size in phi units requires no transformation, and a percentage (of organic content) which varies over a wide range receives the arcsin-square root transformation. Scatter plots and histograms of the residuals would support these choices of transformations.

The observed mean values for each variable, overall, and also by time, by area, and by area-time combination, are given in Table 4.2. These mean values have been back-transformed to the original units of measurement. It may be verified that they closely approximate the population means determined by the simulation criteria as previously described.

The multivariate ANOVA is analogous to the univariate ANOVA of the same design, as can be seen in Table 4.3. The multivariate equivalent of the mean squared deviations column is a column of variance-covariance matrices, and it is obtained in the same way—by dividing through the sum of squared deviations column values by the corresponding degrees of freedom.

Tests of significance in the multivariate ANOVA are based on the sum of squared deviations matrices, rather than on the mean squared deviations matrices as would be analogous to the univariate F-test. We first test the null hypothesis H_{K_1}: "there is no among-location, or patchiness, variation

Table 4.2 Observed mean values back-transformed to original units of measurement; each vector = $[S_1\ S_2\ S_3 : E_1\ E_2\ E_3]$

Time	Control Area	Impact Area	Both Areas
Before impact	[82 41 28 : 4.96 2.33 6.9]	[92 34 22 : 8.4 1.95 6.1]	[87 37 25 : 6.6 2.14 6.5]
After impact	[99 48 24 : 8.64 2.44 6.5]	[148 12 20 : 47.5 2.43 4.9]	[121 24 22 : 25.5 2.44 5.6]
Both times	[90 44 26 : 6.68 2.39 6.7]	[117 20 21 : 25.3 2.19 1.7]	[103 30 24 : 14.8 2.29 6.1]

Table 4.3 Multivariate analysis of variance

Source	df	Sum of Squared Deviations and Cross Products Matrix

Times, $t - 1 = 1$:

$$H_T = \begin{bmatrix} 0.9810 & -1.3318 & -0.3033 & 0.8033 & 0.8898 & -0.4282 \\ & 1.8080 & 0.4117 & -1.0906 & -1.2079 & 0.5813 \\ & & 0.0938 & -0.2483 & -0.2751 & 0.1324 \\ & & & 0.6579 & 0.7286 & -0.3884 \\ & & & & 0.8070 & -0.3506 \\ & & & & & 0.1869 \end{bmatrix}$$

Areas, $a - 1 = 1$:

$$H_A = \begin{bmatrix} 0.6332 & -1.9032 & -0.5193 & 0.6341 & -0.4708 & -0.4778 \\ & 5.7208 & 1.5609 & -1.9050 & 1.4152 & 1.4362 \\ & & 0.4259 & -0.5200 & 0.3861 & 0.3919 \\ & & & 0.6350 & -0.4715 & 0.3553 \\ & & & & 0.3501 & -0.4785 \\ & & & & & 0.3606 \end{bmatrix}$$

Time × Area interaction, $(t - 1)(a - 1) = 1$:

$$H_I = \begin{bmatrix} 0.1823 & -0.7666 & 0.0495 & 0.2519 & 0.2370 & -0.0998 \\ & 3.2234 & -0.2081 & -1.0591 & -0.9964 & 0.4199 \\ & & 0.0134 & 0.0684 & 0.0643 & -0.0271 \\ & & & 0.3480 & 0.3274 & -0.1298 \\ & & & & 0.3080 & -0.1380 \\ & & & & & 0.0547 \end{bmatrix}$$

Locations $\quad H_K$

$$\begin{bmatrix} 2.5753 & -0.0080 & 0.8140 \\ & 3.0188 & -1.0167 \\ & & 2.4537 \end{bmatrix} \begin{bmatrix} -0.4729 & -1.9296 & 0.7876 \\ 0.3095 & 2.7074 & -0.3904 \\ -0.4968 & -1.6699 & 0.7445 \end{bmatrix}$$

$$H_K = \begin{bmatrix} 0.2081 & 0.7872 & -1.0774 \\ & 4.9708 & -0.3164 \\ & & 0.5214 \end{bmatrix}$$

$$at(k-1) = 8$$

Error

$$\begin{bmatrix} 2.6223 & 2.3707 & 1.0717 \\ & 5.5661 & 2.2466 \\ & & 5.2194 \end{bmatrix} \begin{bmatrix} -0.2365 & 0.2728 & 0.4895 \\ -0.7151 & 0.5483 & 1.3136 \\ -0.2648 & -0.2822 & -0.0225 \end{bmatrix}$$

$$E = \begin{bmatrix} 0.7082 & -0.1594 & 0.2737 \\ & 2.1444 & -1.0424 \\ & & 2.0651 \end{bmatrix}$$

$$atk(r-1) = 24$$

Total

$$\begin{bmatrix} 6.9941 & -1.6389 & 1.1126 \\ & 19.3372 & 2.9945 \\ & & 8.2062 \end{bmatrix} \begin{bmatrix} 0.9799 & -1.0009 & 0.2713 \\ -4.4613 & 2.4665 & 3.3607 \\ -1.4616 & -1.7767 & 1.2191 \end{bmatrix}$$

$$T = \begin{bmatrix} 2.5571 & 1.2123 & -0.9666 \\ & 8.5803 & -2.3259 \\ & & 3.1886 \end{bmatrix}$$

$$atkr - 1 = 35$$

in the biotic variables." If this hypothesis is rejected, patchiness must be considered to be a significant source of variation. The appropriate denominator for testing the null hypothesis H_{I_1}: "there is no areas-by-times interaction for the biotic variables" would then be the among-locations matrix H_{K_1}, which is the upper-left submatrix of the H_K matrix in Table 4.3. The numerator would of course be the matrix H_{I_1}, which is the upper-left submatrix of the H_I matrix.

If the null hypothesis H_{K_1} were accepted, it could be assumed that no significant patchiness variation was present, and the among-location variation could be pooled with the within-location variation for use as the denominator in testing the hypothesis H_{I_1}. Matrices H_{K_1} and E_1 in Table 4.3 would be summed, term by term, and the resulting matrix $H_{K_1} + E_1$ (with $8 + 24 = 32$ df) would represent the pooled "within time-by-area combinations" variability.

Various test statistics are calculated by the SAS procedure, including Pillai's Trace V which is recommended by Olson (1976) as the least likely of all the commonly used test statistics to produce a Type I error (Section 2.1.3) and the least affected by violations of the assumptions such as heterogeneity of variance-covariance matrices. For the test of the null hypothesis H_{K_1}: "no among-location variation in the biotic variables" $V = 1.3000$ and the F-approximation is $F(24, 72 \text{ df}) = 2.29$ ($p < 0.01$). Therefore, we reject the null hypothesis, conclude that there is significant patchiness variation, and use matrix H_{K_1} as the denominator for the test of the null hypothesis H_{I_1}. For this test $V = 0.5736$ and $F(3, 6 \text{ df}) = 2.69$ ($p = 0.14$). If these were biological field data, we would accept H_{I_1} and conclude that there has been no change in the biotic community of the impact area that did not also occur in the control area. However, we know that these data were simulated to have such a change, and we must conclude that either more locations per area-by-time combination or more replicate samples per location were required for detection of the area-by-time interaction. In this case we know that the interaction is really present and proceed to describe it.

Because there is only one degree of freedom for interaction there can be only one canonical variate associated with species interaction. The canonical variate associated with the ratio of the H_{I_1} matrix to the H_{K_1} matrix is CVSI $= 0.23 \ln(N_1 + 1) - 0.97 \ln(N_2 + 1) + 0.06 \ln(N_3 + 1)$, or any equation with coefficients proportional to these. A canonical variate score may be calculated for each of the 36 samples by substituting the observed species abundances N_j into this equation; these 36 scores will then represent values of a new variable that maximizes the ratio of areas-by-times interaction variation to among-location, or patchiness, variation. The times-by-areas matrix of CVSI score means is

$$\begin{array}{c} B \\ A \end{array} \begin{bmatrix} -2.40 & -2.21 \\ -2.52 & -1.15 \end{bmatrix}$$
$$\quad\quad C \quad\quad I$$

which shows a sharp contrast between the CVSI score mean for the impact area after impact (AI) on the one hand and the control area means (AC and BC) and the mean for the impact area before impact (BI), on the other hand.

The CVSI vector coefficients may be interpreted as relative contributions by the variables to the areas-by-times interaction. The interpretation here is that the change unique to the impact area involves an increase in abundance of species S_1, a large decrease of S_2, and relatively little change in S_3. Any attempt to attach significance to the magnitudes of these coefficients or to select a significant subset by some stepwise procedure would be inappropriate, given the purpose of our analysis, as well as analytically invalid with such correlated variables (Section 3.6). However, the relative magnitudes are consistent with the species abundance changes on impact that went into the data simulation.

It may be noted that we are interpreting the vector coefficients as relative contributions without having first standardized them. This is not necessary if standardized data are used in the analysis, and the effect of the logarithmic transformation is to accomplish just that (Section 2.3.9).

Now we repeat the analysis for the environmental variables, working with the lower-right submatrices in Table 4.3. The test of the null hypothesis H_{K_2}: "there is no among-location variation in the environmental variables" yields an F-approximation based on the test statistic V of F (24, 72 df) = 1.27 ($p - 0.22$). Here we could conclude that there is no significant patchiness variation in the environmental variables, and pool matrices H_{K_2} and E_2. However, we know that these data are simulated to have such patchiness variation, and must conclude that the number of locations and replicate samples were insufficient to detect it. Also, we wish to keep the statistical design comparable for the biological and environmental data. Therefore, we use matrix H_{K_2} as the denominator for the test of the null hypothesis H_{I_2}. For this test F (3, 6 df) = 41.2 ($p <$ 0.01), and we conclude that there was an environmental change unique to the impact area. The canonical variate associated with this environmental areas-by-times interaction is CVEI = 0.66 arcsin $\sqrt{E_1/100} + 0.70E_2$ $- 0.28 \ln E_3$. Variable E_1 is not log-transformed, and therefore interpretation of the three coefficients as relative contributions is questionable. The signs are, however, consistent with the properties of the simulated data in that increased organic content (positively correlated with fineness of sediment) is associated with decreased oxygen concentration. Again, canonical variate scores may be calculated for each of the 36 samples and the times-by-areas matrix of CVEI score means is

$$\begin{matrix} B \\ A \end{matrix} \begin{bmatrix} 1.24 & 1.05 \\ 1.39 & 1.76 \end{bmatrix} ,$$
$$ C \qquad I$$

with a high value for the impact area after impact.

We now wish to describe the relationship between the biotic areas-by-times interaction and the environmental areas-by-times interaction. There is only one degree of freedom associated with the interaction and therefore only one dimension's worth of variation in the interaction submatrices H_{I_1} and H_{I_2}. One variable suffices to represent that variation in each case, and the canonical variates CVSI and CVEI do exactly that. If the interaction is caused only by the impact effects (which should be true unless some change unrelated to impact has occurred in one area but not in the other), the CV vector coefficients describe the relationships having to do with impact between the biotic variables and the environmental variables. When only one degree of freedom is associated with the areas-by-times interaction there is no basis for additional formal tests. The H_I matrix may be converted to a correlation matrix by dividing the off-diagonal values by the square roots of the two corresponding diagonal values; if this is done all correlations can be seen to be plus or minus unity. This is not true for matrices such as H_K, which are associated with more than one degree of freedom.

However, if the control and impact areas and the before and after times have been properly selected, we can assume that changes unique to the impact area were caused entirely by the effects of the impact. It therefore will follow that the CVSI and CVEI vectors describe a single impact-related phenomenon. The importance of careful selection of areas and times should be obvious. Biotic and/or environmental similarity of areas before impact, and of times for the control area, can be tested. The tests are described and illustrated later. That any change unique to the impact area is impact-related can only be assumed. If there were more than two areas (e.g. − high, low, and zero impact areas), there would be additional interaction degrees of freedom for testing and interpretation. This is discussed later.

A scatter plot of CVSI versus CVEI scores, as shown in Figure 4.1, effectively displays the results and provides a basis for biological monitoring criteria to be used for detection of such impacts in the future. The latter is described in Section 4.3. In Figure 4.1 each letter A, B, ..., L, represents a location, and each consecutive set of three letter (A–C, D–F, G–I, J–L) represents an area-by-time combination. Letters G–I represent

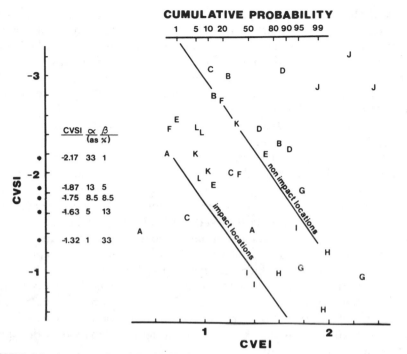

FIGURE 4.1 A scatter plot of the CVSI (Canonical Variate for Species Interaction) scores versus the CVEI (Canonical Variate for Environmental Interaction) scores. Each letter A–L represents a sampling location. Letters G–I represent impacted locations and all other letters represent nonimpacted locations. The parallel lines describe cumulative percentage plots on an arithmetic probability scale (at top of figure, as percentages). At the left are given several critical CVSI values as possible monitoring criteria, with their α and β error levels.

the impact area after impact. Observe that despite substantial among-replicate variation within locations (same letter variation) there is also variation among locations within areas-by-times combinations. That is, same letters tend to cluster. The 9 samples from the impact area after impact are clustered. Except for the among-locations variation, the other 27 samples are randomly intermixed. That is, there is no evidence that areas-by-times combinations (the sets $A-C$, $D-F$ and $J-L$) tend to cluster.

If the null hypothesis H_{I_1} is true, the tests for main effects (hypotheses H_{T_1} and H_{A_1}) are meaningful tests for change between times in both areas, and difference between areas at both times, respectively. However this is not so if H_{I_1} is false, as can be seen from the mean values in Table 4.2. Significant main effects may be a byproduct of a significant interaction, as would be true in this example. The appropriate test for change between times is then based on the control area only, and the appropriate test for difference between areas is based on the before-impact time only. In both tests there is a multivariate ANOVA design with three locations nested in each of two groups. There are 18 samples. The test statistic V is calculated for the hypothesis that the two groups (times in one test and areas in the other) do not differ. In both tests we obtain F's with $p >> 0.05$. The null hypothesis is accepted in both cases, which is in agreement with the criteria for the data simulation.

The null hypothesis H_0: "no areas-by-times interaction for the biotic variables" was correctly rejected when it was false. Would it be correctly accepted if it were true? The simulated data are now modified by removing the changes in the biotic variables due to impact, so that mean abundances of S_1, S_2, and S_3 do *not* change by $+50$, -75, and -30 percent respectively after impact in the impact area. Otherwise the simulated data remain the same, with the same standard deviations and intervariable correlations and the same among-location patchiness. The four areas-by-times vectors of S_j mean values remain the same as given in Table 4.2, except for the after-impact (AI) vector which becomes [99 47 29] rather than [148 12 20]. If the same factorial multivariate ANOVA with nested locations is now repeated, the test of the hypothesis H_{I_1} yields an F with $p >> 0.05$. Thus we do accept the true null hypothesis. As would be expected, the hypothesis H_{K_1} of no patchiness variation among locations yields the same significant F-value as before.

We now wish to effectively display the results to make the effects of impact stand out by using an analysis approach different from the factorial multivariate ANOVA approach used previously. Both philosophical and methodological choices are involved here. Methodologically one might generate an effective display from an ordination or a cluster analysis (see Section 4.4), from spanning trees, contour plots, or any number of techniques used to effectively display multivariate data (Section 3.11). From the many kinds of cluster analyses I choose for this analysis an agglomerative hierarchical cluster analysis method based on Orloci's (1968) measure of mutual information. To display the results I use a perspective view of the cluster analysis dendrogram (or tree). Program listings are given by Orloci (1975a).

The philosophical choice is independent of the analysis and display methods used. One must ask, Is it more important that the method show the impact effects as clearly as possible or that it be independent of the previous analysis so that any evidence of impact effects will corroborate the conclusion that there *are* impact effects? If, for example, effective display of impact were more important, one should cluster from the CVSI score vector which by definition most efficiently describes the areas-by-times interaction, assumed to be entirely impact related. Williams and

Stephenson (1973) suggest the possibility of clustering from a multivariate ANOVA interaction vector. Macnaughton-Smith (1963) discusses clustering to optimize an external criterion. However, in reality such an approach only provides a different display format for the same statistical analysis results and cannot in any way corroborate the previous conclusion. The other philosophical choice, which is the one I make here, is to perform an analysis of the biotic data that does not depend in any way on the analysis used previously. Impact effects will not be optimally displayed, but if they *are* detectable they increase the robustness of the conclusion that there are impact effects (Sections 2.1.6, 2.1.7).

The display of the cluster analysis results is in Figure 4.2. The height at which any two samples or groups of samples are fused is equal to the

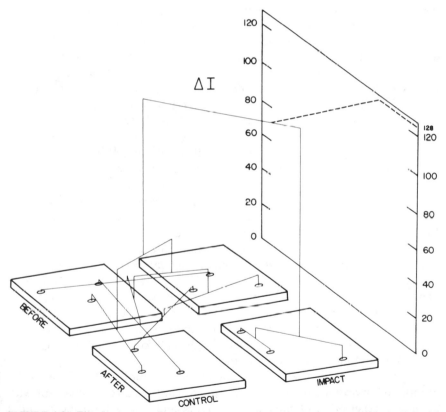

FIGURE 4.2   The dendrogram from the cluster analysis on species compositions of the 36 samples. The ordinate (ΔI) is the gain of information on fusion of any two samples or groups of samples. Fusion at higher levels represents greater dissimilarity in species composition between the samples being fused.

gain in information ($\triangle I$) on fusion. If the samples being fused have identical species compositions then $\triangle I$ is zero. The most biotically similar samples fuse first at the lowest levels, and the groups of samples that fuse last (and highest) are most dissimilar in species composition. The last fusion connects the 9 impacted samples to the remaining 27 nonimpacted samples, and the fusion occurs at a much higher level than all previous fusions. Furthermore, there is no pattern in the previous fusions. Samples within the same area-by-time combination show no tendency to fuse at lower levels with connections to other areas-by-times combinations at higher levels. The impacted samples form a group, or cluster, with a similar species composition, which is different from that of the nonimpacted samples. Again we conclude that there are impact effects on the biota.

The sampling and analysis design presented here is an optimal design qualitatively. In a particular impact study choices must be made about how many samples, locations, variables, areas, and times should be used. "The more the better" is not generally true and even when true the combined effects of cost and the law of diminishing returns would require that choices be made. See Sections 2.3.8, 3.6 and, 3.8 and earlier parts of this section for discussion. Necessary sample number for detection of a specified change, based on data from preliminary sampling, could be estimated as for the univariate factorial ANOVA design (Section 2.38). However, the number of possible ways the H_0 can be violated (i.e., the number of possible H_A's) increases rapidly with the number of variables, so that the specification of exactly *what* change must be detected becomes rather complicated. Simulation of data sets for specified H_A's would be a possible approach. For number of replicate samples the more the better probably *is* the best rule of thumb, because of the dependence of the robustness of multivariate tests on large H_0 degrees of freedom. The choice of number of locations depends on the among-location patchiness relative to the within-location variability. Assume that the total number of samples per area is to be 24. At one extreme, where there is almost the same amount of variation among samples at the same location as among different locations within the area, one might scrap nested locations entirely and allocate 24 samples randomly over each area at each time. At the other extreme, a very patchy environment might call for 3 random samples at each of 8 randomly allocated locations.

The best rule to follow for both the number of biotic variables and the number of environmental variables is the fewer the better, consistent with adequate description of the impact effects and any natural background variation. Selection procedures and criteria have been discussed (Section 3.6).

If some or all of the variables are nonquantitative, there are several possible approaches. If all variables are binary or other category variables, multidimensional contingency table analysis as described in Section 3.4.1 could be appropriate. Alternatively, binary or rank data could be used, preferably after application of the Fisher-Yates transform, in the quantitative analyses as described here (Sections 2.3.9, 3.4.1, 3.4.2). Many cluster analysis procedures are available for binary or mixed data. However, the recommended approach, which should efficiently handle all data types or mixtures of them, is to first perform a principal components or coordinates analysis as described and illustrated by examples in Section 3.4.3. This will convert the observations on the original variables to observations on a new set of uncorrelated quantitative variables, which can be then used in an analysis such as the one just presented.

The use of two areas, one a control area and the other an impact area, has a certain logical neatness that simplifies the interpretation of results. However there is one substantial advantage to the use of several areas representing different degrees of impact—the number of degrees of freedom associated with the areas-by-times interaction will then be greater than 1. More than one canonical variate can therefore be calculated for the biotic and for the environmental areas-by-times interaction, and this opens up new possibilities. Assume, for example, that there are three areas (high, low, and zero impact). With two times, the interaction degrees of freedom will be $(a - 1)(t - 1) = (2)(1) = 2$. If the three areas have been properly selected, the first of the two possible canonical variates for interaction should represent almost all of the interaction variation and the second should be trivially small. If this is not the case, more than one phenomenon is probably involved, which must mean that the areas chosen are influenced by some change other than impact effects.

It is best that there be either two times or many. Many implies time series analysis as discussed previously and in Section 4.2. With two times, the before-impact time should be as short as possible to minimize the chance that a nonimpact change with differential effect in the two areas might take place. The after-impact time must be chosen with regard to the type of impact involved and the response pattern of the biological community to it (see Section 3.8).

Finally, the reader should refer back to Section 3.2 and judge for himself or herself whether the example and possible modifications of it, as presented in this section, satisfy the criteria for an optimal impact study design.

4.2   MAIN SEQUENCE 2: IMPACT IS INFERRED FROM TEMPORAL CHANGE ALONE

Here the sampling and statistical analysis design is suboptimal in that a control area defined *a priori* is lacking. The options are those in rows 1 and 2 and columns 3 to 6 of Figure 3.2. Sampling at different times may be at a number of different locations (row 2), in which case temporal change in spatial pattern can be the criterion of impact effects. If there is only one location (row 1), time-series analysis modeling in the strict sense must be used. In most cases the statistical analysis model used to relate biological criterion variables to environmental predictor variables is that of the second row in Figure 3.3.

Green (1977) describes a sampling design and statistical analysis approach for the situation where samples are allocated spatially at locations on a grid, and temporally once before and once after impact as in the example of Section 4.1. Assume that an environmental change whose time and place of occurrence are known may—or may not—affect the biological community within some defined area that has an existing pattern of natural environmental variation and a distribution of abundance of species related to that pattern. The null hypothesis is that the species abundances are not affected by the environmental change, and if it is rejected, we wish to describe how change in the species abundance variables is related to change in the environmental variables. The example is based on simulated data, with the following properties.

1. The relative abundances of six species and their distributions on four environmental variables were chosen so as to be similar to values from benthic communities in Manitoba lakes. These properties are summarized in Table 4.4.

2. The simulated environment over which these species are distributed is that of a portion of a lake basin similar to a part of a real lake—the south basin of Lake Winnipeg. The environmental change that is simulated has the potential of being a real one—an increased load of salts, which could enter Lake Winnipeg via the Red River as a result of the Garrison Diversion project. The four environmental variables are depth (in metres), sediment mean particle size (in phi units), sediment percentage organic content, and calcium concentration in the bottom water [as milligrams per litre of calcium carbonate $(CaCO_3)$]. Ninety-two stations are on a kilometre grid and are sampled at two times, once before and once after the calcium concentration of the incoming water at the south end of the basin has increased by 60 mg/l. Figure 4.3 shows the spatial pattern for depth, and Figure 4.4 shows the calcium concentration at both times. Values for the sediment variables were randomly sampled from the range of values observed for the given depth and wind exposure in Lake Winnipeg.

3. The species abundance data were generated from normal deviates with relative abundances, means, and standard deviations for each species appropriate to the value of each environmental variable at the given station

Table 4.4 Relative abundances and distributions of species on environmental variables used for the generation of artificial data

Species	A	B	C	D	E	F
Relative abundance	100	84	84	79	58	52
Mean (standard deviation as % of mean)						
Calcium (mg/l as $CaCO_3$)	60 (22)	99 (7)	99 (15)	81 (16)	122 (6)	99 (9)
Depth (m)	2.4 (32)	8.2 (33)	3.0 (27)	5.0 (35)	4.5 (27)	5.0 (38)
Mean particle size (phi)	1.0 (110)	1.6 (81)	1.1 (127)	1.7 (88)	2.3 (61)	2.1 (57)
% organic content	7.5 (83)	15.0 (60)	6.9 (59)	10.0 (60)	11.0 (55)	19.0 (58)

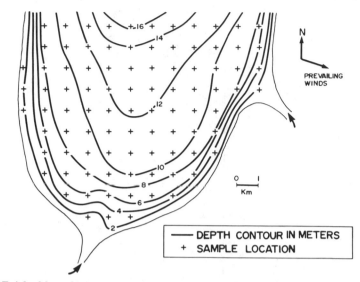

FIGURE 4.3 Map of lake basin with depth contours and sample locations. Figures 4.3 to 4.7 are produced by permission of the American Society for Testing and Materials from figures of Green (1977).

and time. Logarithms of species abundances were generated from logarithms of environmental variables (sediment particle size in phi units is already in logarithmic form), to better approximate the nonnormal distributions and nonlinear relationships characteristic of such data.

4. In order to include the noise caused by sampling error that is characteristic of such data, each value of logarithm of species abundance was replaced by a random normal deviate with mean equal to that value and standard deviation equal to 50 percent of that value. The antilog of this randomly perturbed value was then used as the species abundance in the sample, rounded to whole numbers of organisms. The resulting species distributions are shown in Figure 4.5.

Now that data with known properties have been created, let us forget that fact for a while and slip into the role of the applied ecologist. The first question to ask of the data is whether there has been, in fact, an environmental change of the type anticipated between times 1 and 2, and whether any other environmental variables changed during that time. A multivariate analysis of variance indicates that the two times differ significantly ($p < 0.01$) in the values of the four environmental variables. Given this overall environmental difference between times, univariate analyses of variance indicate that the only variable of the four that differs

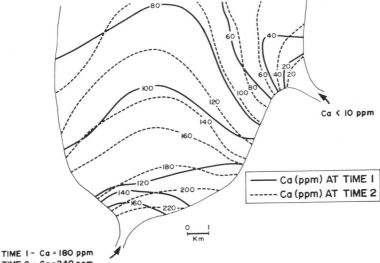

FIGURE 4.4 Map of lake basin showing calcium concentration at two times.

significantly is calcium. This result agrees with the known properties of the data.

Multivariate analysis of variance fails to reject the null hypothesis that the two times are similar in overall abundances of the six species ($p > 0.05$). This is a crude test of whether the biological community has changed, however. The variation among stations within each time is high because of the influence of depth and sediment type on the distributions of species. Also, any change in species abundances between times would be expressed more as a shift of pattern than as a change in overall abundance.

What we wish to test and describe is the relationship between biological change and environmental change. A model that assumes a linear additive relationship between species abundances and environmental variables is inappropriate (see Section 4.4). However, a model that assumes a linear additive relationship between changes in species abundances and changes in environmental variables, over short time intervals, is appropriate. All data would, as usual, be transformed logarithmically, and in this model the effect would be to imply percentage change rates that are linearly and additively related. The spatial, within-time variation in both biological and environmental variables is thereby removed from the analysis. This strategy is analogous to the use of a paired t-test for comparing the means of two populations where there is extreme within-group variation and a log-

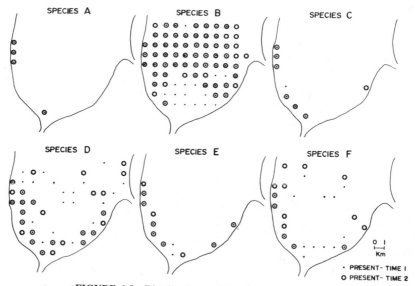

FIGURE 4.5 Distributions of six species at two times.

ical pairing of individuals between the groups. In our data, the stations pair across time (see Section 4.1).

The overall regression analysis yields one significant ($p < 0.01$) canonical variate:

$$-0.14 \, \triangle \log A + 0.04 \, \triangle \log B + 0.02 \, \triangle \log C - 0.04 \, \triangle \log D - 0.10 \, \triangle \log E$$

$$+0.18 \, \triangle \log F = -0.86 \, \triangle \log \text{Ca} + 0.09 \, \triangle \text{phi} + 0.06 \, \triangle \log \% \text{ organic}.$$

Depth is omitted because the stations are fixed, and there is no change in depth between times. Changes in particle size and percentage organic content of sediment reflect the small-scale patchiness of the substrate at given stations and are a result of sampling variation between times. The relationship between the changes in species abundances and the changes in the environmental variables is shown by the results of a canonical correlation analysis in Figure 4.6. The null hypothesis that changes in species abundance are unrelated to changes in calcium is rejected ($p < 0.01$). The null hypothesis that changes in sediment mean particle size contribute nothing additional, beyond what changes in calcium contribute, to changes in species abundance is rejected ($p < 0.05$). There is no additional contribution by percentage organic content ($p > 0.05$).

The pattern of change between times in each of the three environmental variables is shown in Figure 4.7. As has been shown, only calcium has a significant overall change between times, and it is apparent that the

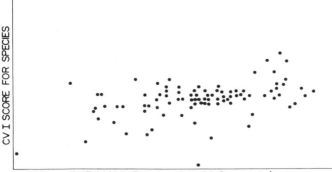

FIGURE 4.6   Canonical variate I scores for changes in six species abundances plotted against canonical variate I scores for changes in three environmental variables.

changes in the sediment parameters are without pattern. The pattern of change in abundance of each of the six species was shown in Figure 4.5.

Figure 4.7 also shows the pattern of change in three additional biological variables: S is the number of species, the diversity index H' is estimated by $-\Sigma_j (N_j/N) \log(N_j/N)$, where N_j is the number of organisms in species j and N is the number of organisms of all species. Diversity indices are often used to reduce species abundance data to a single biological variable for each sample (see Section 3.5.2). The index H', which takes into account both species number (or richness) and equitability of distribution of number of organisms among species, is often used with quantitative data. These data offer the opportunity to determine how much information is lost by reduction from abundance changes for six species, to changes in S, H', and N.

The overall regression analysis yields a canonical variate that just fails significance ($p = 0.07$):

$$0.10 \,\triangle S - 0.04 \,\triangle H' - 0.00 \,\triangle N = -0.86 \,\triangle \log Ca + 0.08 \,\triangle phi -$$

$$0.02 \,\triangle \log \% \text{ organic.}$$

The null hypothesis that changes in these biological variables are unrelated to changes in calcium is rejected ($p < 0.05$). Neither sediment variable is significant. The simple linear regression of change in S on change in calcium is significant at $p < 0.01$, and the regression of change in H' on change in calcium is significant at $p < 0.05$. The regression of change in N on change in calcium is not significant ($p > 0.05$).

It is interesting that change in S is a better measure of the biological change, with respect to its relationship with the change in environment,

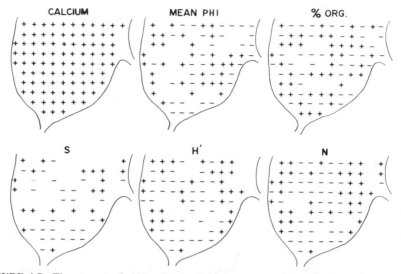

FIGURE 4.7 The pattern of change between times in three environmental variables and three biological variables. The symbol + represents increase between times 1 and 2, and − represents decrease.

than is change in H'. This result supports Hurlbert (1971) who argues that such information or entropy-based indices as H' have neither obvious biological meaning nor demonstrated empirical value. With computers there is no greater difficulty in using the species abundance changes as dependent variables than in using S or H' change as a single dependent variable. Since all the biological information is retained with the former, it is preferred.

All analyses (including the graphic displays) were carried out using the SAS computer package (see Section 3.10).

If sampling at all stations is repeated more than two times (row 2, columns 4 to 6 of Figure 3.2), there is no problem. In fact such data will allow more sensitive tests of hypotheses. One approach would be to simply calculate the average change in each variable (biological and environmental) per unit time at each station, and use this instead of the difference between the two times. Or, if the times of sampling cover prechange equilibrium values, the period of change, and postchange equilibrium values, a logistic or cumulative normal model could be fitted for each variable and station to estimate the before-impact and after-impact asymptotic values of the variables for each station. Then the differences between these before and after levels could be used exactly as described above.

Williams et al (1969) consider analysis approaches to change in spatial pattern of species composition over a series of times, and apply ordination, clustering, and diversity indices to forest succession data. Their most effective procedure begins with the use of cluster analysis on species presence-absence data (118 species) for 120 location-times (10 locations on a rectangular grid and 12 times over a 7 year period) to produce seven groups of location-times with differing species compositions. An effective temporal-by-spatial display of the results is possible, as shown in Figure 4.8. It is apparent that early succession is characterized by temporal shifts in species composition that occur at all locations, whereas in later succes-sional stages the heterogeneity of the study area becomes evident in the largely spatial definition of the species groups. Such an analysis and display could be applied effectively to a situation where an impact occurred at a particular time and location and its effects on the biotic community persisted for some time and perhaps changed spatially as well. Where times were over cycles (seasonal, diurnal) polar graph paper might be used effectively. Also, these species groups as biological criterion vari-ables could be related to environmental predictor variables using discrim-inant analysis as described in Section 4.4. Williams et al go on to describe the use of transition matrices (Usher 1972, Davis 1973, Poole 1974, Harris 1975) to model succession, based on these species groups. Dale et al (1970) carry this development of "numerical classification of sequences" further and point out an error in the treatment by Williams et al.

A standard reference on time series analysis is by T. W. Anderson (1971). Simulated and "real" data given in an appendix are used for examples. In an earlier paper Anderson (1963) reviews the state of the art in both univariate and multivariate time-series analysis, noting that many of the advances in this area have come from economics. They also have come from geology, with its interest in stratigraphy and the succession of paleocommunities in paleoenvironments. Chapter 5 of Davis (1973) is an excellent basic reference, with geological examples, computer program listings, and result display formats. Two other basic references are by Munn (1970) and Batschelet (1976), and the latter is available in paperback. Extensions to multivariate time series, on a more technical level, are by Quenouille (1957) and Hannan (1970).

For time-series analysis with binary data the standard reference is Cox and Lewis (1966). See also Davis (1973). As an example consider that one has species counts over a series of times and wishes to test against H_0: "no variation in species success (as presence) among times." Cochran's (1950) Q test, which is well treated by Siegel (1956), is appropriate. The data in Table 4.5 are for species of desmid phytoplankton collected from Mountain Lake, Virginia, between May 16 and December 9, 1970, by Oberg-Asamoa and Parker (1972), and are taken from their Table III. The

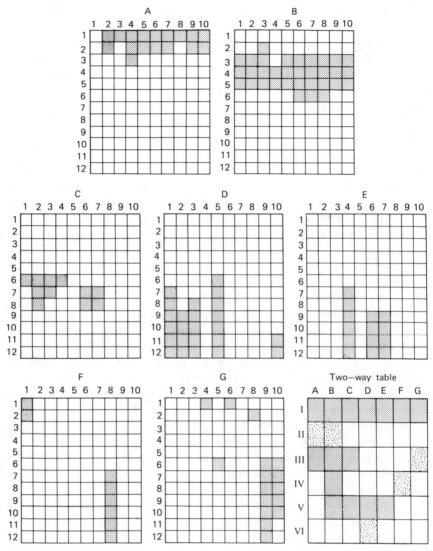

FIGURE 4.8 Seven species assemblage sample groups (A–G) derived from a cluster analysis on rain forest species successional data, plotted on a grid of 12 times by 10 locations. The "two-way table" plots characteristic floristic groups against the sample groups. Reproduced by permission of the British Ecological Society from Figure 6 of Williams et al (1969).

Table 4.5 Species of desmid phytoplankton collected from Mountain Lake, Virginia, between May 16 and December 9 of 1970 (from Obeng-Asamoa and Parker 1972, Table III)

Times	1	2	3	4	5	6	7	8	9	10	11	12	13	14	15	16	17	18	19	20	21	22	23	24	25	G_j	G_j^2
1																	1	1								2	4
2																			1							1	1
3	1							1			1		1	1							1	1	1			8	64
4	1	1	1		1		1		1	1	1	1		1		1	1			1	1	1	1	1	1	18	324
5	1	1	1		1				1	1	1	1				1				1	1	1	1		1	14	196
6	1		1		1				1	1	1	1				1				1	1	1	1		1	13	169
7	1	1							1	1	1	1								1	1	1	1	1	1	12	144
8	1		1						1	1	1	1								1	1	1	1	1	1	12	144
9	1				1	1	1				1									1	1	1	1		1	10	100
10	1				1		1		1	1	1	1							1	1	1	1	1		1	13	169
11	1			1					1	1	1	1								1	1	1	1		1	11	121
12											1				1	1						1	1			5	25
L_i	9	3	4	1	5	1	3	1	7	7	10	7	1	2	1	4	2	1	2	8	9	10	10	3	8	119	
L_i^2	81	9	16	1	25	1	9	1	49	49	100	49	1	4	1	16	4	1	4	64	81	100	100	9	64		

$\sum_i L_i^2 = 839 \qquad \sum_j L_j^2 = 1461$

H_0: "no change in species composition over a season" is almost certain to be false. Therefore, this is a rather trivial example but it will serve to illustrate the test. If the statistical H_0: "the ones are randomly distributed in the rows and columns of Table 4.5" is true, then

$$Q = \frac{(k - 1) [k \sum_j G_j^2 - (\sum_j G_j)^2]}{k \sum_i L_i - \sum_i L_i^2}$$

where there are $j = 1, k$ rows and $i = 1, p$ columns. The sums for rows and columns are, respectively, G_j and L_i. The sampling distribution of Q is approximated by X^2 with $k - 1$ df, if the number of columns (species) is not too small. For our example

$$Q \ (11 \ df) = \frac{(11) [(12) (1461) - (119)^2]}{12(119) - 839}$$

$$= 63.$$

Under H_0 a value as large as X^2 (11 df) = 63 or larger has a probability of $p < 0.01$. We reject H_0, concluding that species diversity as measured by the number of species S (see Section 3.5.2) does vary with season.

In a real impact study with a larger data set this rejection of H_0: "nothing is going on in this temporal sequence of binary species data" (see Section 3.4.3) might be followed by clustering and ordination to effectively describe the temporal structure of the data. For example, principal coordinates analysis clustering (see Section 3.4.3), which would be particularly appropriate where the number of species greatly exceeded the number of times, would allow the display of temporal species-groups in a most efficiently reduced space. The temporal species-groups, or clusters, could be connected by arrows to indicate the direction of movement along the path through the ordination space. Clustering could also be done by Williams and Lambert's (1959, 1960) association analysis (Sections 3.4.1, 3.6) which would identify best indicator species. The same sequence of hypothesis testing, descriptive analysis, and effective display could be followed using quantitative species data, although the specific tests and analyses would differ.

In addition to the general time series references previously given, Wald and Wolfowitz (1943) and Hannan (1960) describe hypothesis testing and estimation of confidence intervals in univariate time series. Jones et al (1970) describe the use of multivariate autoregression analysis for building a sufficient model to describe preevent data, and then testing departure from that model in postevent data. Jensen (1972) describes simultaneous multivariate procedures for testing stability of a p-variate vector over time, for testing stability up to some time and after that time, and for testing stability at the same times in two areas. Two FORTRAN programs by Bower et al (1974) model baseline data that may contain cyclic and noncyclic trends, and then analyze intervention effects given the baseline model.

Seasonal and other cyclical effects are likely to be important in many environmental studies, and must be incorporated into baseline models. Livingston (1976, 1977), for example, considers diurnal, tidal and seasonal cycles in estuarine systems as they relate to pollution studies. Techniques for estimation of seasonal and other periodic components in time series are described by Durbin (1963) and Bulmer (1974).

Correlations between time series, say between biological and environmental variables, could indicate causal relationships. However, they could also result from responses to the same unmeasured natural rhythms, so that great caution in interpretation of such correlations is necessary. Slobodkin (1968) notes that correlation of time series is usually the wrong approach for detecting relationships between species populations, which— if they are at or near equilibrium—will show no statistical correlation. Hamon and Hannan (1963) discuss the estimation of relations between time series, and Saila et al (1972) provide an example of time-series analysis used to correlate alewife activity and environmental variables at a fishway. Two other examples of the application of time-series analysis are by Fox (1969) who models and simulates paleoecologic communities through time, and by Bulmer (1974) who presents a statistical analysis of the 10-year cycle in Canadian mammals.

For computer programs applicable to time-series analysis see BMD (Dixon 1973), Davis (1973), Bower et al (1974), and SAS (Barr et al 1976).

Toxicological studies can be thought of as a type of time-series analysis where the biological criterion variable "survival" is related to controlled levels of an environmental predictor variable that is usually the agent of impact. Such studies can be done either in the laboratory or the field, but a characteristic is always the ability to control both the levels of the agent and the homogeneity of the biological response material. Total reliance on descriptive field studies, even with spatial and temporal controls, is often inadequate. Uncontrolled environmental variables of various kinds influence the biological responses in the field, and the testing of potentially harmful agents before impact is desirable. Even if one were willing to take a chance and see what happens, the time is often not available for adequate baseline studies. Waldichuk (1973) argues for bioassay techniques and remarks that baseline field studies require a minimum of 2 years, better yet 5 years, and best of all 10 years. Evaluating the effects of a wide range of levels of the agent over different lengths of time on a variety of organisms is often the best approach. Best of all, of course, is controlled toxicological studies done hand-in-hand with field studies of the kind described previously.

The classic paper by Sprague (1969) is a good basic reference. He reviews earlier work such as the formal probit analysis methods of Finney (1964), provides step-by-step description and illustration of toxicological study procedures, and discusses various analysis and display options. Sprague describes three main steps. First, replicate sets of organisms should be exposed to each of a series of concentrations, or levels, of the agent and the numbers still alive at each of a series of times should be recorded. The mean or median survival times are then estimated. Second, plot concentrations against the logarithm of mean or median survival times on graph paper (see Brett 1956). Any nonlinearity of the line would suggest unusual features of the toxic action. The point at which the line levels off is the lethal threshold concentration. Third, estimate the LC_{50} (lethal concentration for 50 percent of the organisms) by choosing an exposure time from the horizontal part of the curve and plotting mortality against concentration on log probit (logarithmic probability) paper. Fit a straight line statistically or by eye, and estimate the LC_{50} value. Sprague emphasizes graphical rather than formal statistical techniques, stating that they are easier and often as good. This argument is perhaps less forceful with the increased availability of computer programs for these methods.

I would urge that the very *first* step with a new organism or agent should always be a preliminary run to estimate the approximate ranges of times and concentrations of the lethal action, and the necessary number of replicate organisms for sufficient precision of the LC_{50} estimate (Sections 2.3.5, 3.8). The shape of the response curve (mortality *or* survival) is sigmoid, asymptotic to 0 percent at one end and to 100 percent at the other. Whether the probit or logit transformation (see below) is used for conversion of this sigmoid curve to a straight line, observed values of 0 or 100 percent mortality do not contribute to estimation of the response curve—hence the importance of allocation of concentrations and times within the lethal range estimated from a preliminary run.

The sigmoid response curve can be adequately approximated by either the cumulative normal or the logistic models, from which are derived the probit and logit transformations respectively. Each has certain theoretical or computational advantages, but the variability in most data is greater than the slight differences in the shapes of the curves. There have been violent probit-versus-logit debates that in retrospect appear to have been rather pointless, at least to the practical environmental biologist. Probit methods (Finney 1964) have seen greater use than logit methods (Berkson 1944, 1946, 1951), but such workers as Colquhoun (1971) conclude that one is about as good as the other.

There are several approaches to significance testing and confidence limits for LC_{50} and related parameters. Sprague (1969) emphasizes simplified methods applicable to median lethal times, and he provides a literature review on the subject. More formal statistical methods for fitting response curves often have associated tests of significance, but a simple approach is to run entire toxicological experiments in replicate and estimate LC_{50} for each. These replicate LC_{50}'s can then be used in statistical tests and for the setting of confidence limits. Green (1965) describes a method for estimating tolerance over indefinite time periods, in which concentrations are plotted against the reciprocal of mean or median survival times (see also references cited by Sprague). An approximately straight line results (see Figure 4.9) and the intercept, which can be estimated with confidence limits by standard linear regression methods, is the estimated concentration at which 50 percent of the organisms die over an indefinite time period. Sprague remarks that confidence limits from toxicological studies are often misinterpreted, in that they are not for the organism or agent in general but rather for experiments conducted in exactly the same way. Action of the agent on organisms in the field may be very different.

See Section 2.3.9 and Figure 2.11 for presentation of Cole's (1962) closed sequential design and its application to toxicological studies where exposure to the agent of impact and to control conditions can be paired sequentially. The need for a control because of the artificial conditions under which most toxicological studies are done was mentioned in Section 2.3.4. Cox (1972b) describes the application of a log-linear regression model to data where survival is a function of both time and a specified

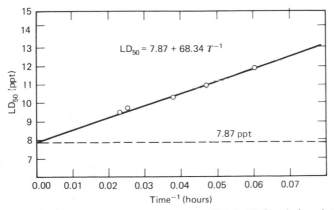

FIGURE 4.9 LD_{50} ($=LC_{50}$) values for *Pontoporeia affinis* in NaCl solution plotted against the reciprocal of the number of hours of exposure. Reproduced with permission from Figure 1 of Green (1965).

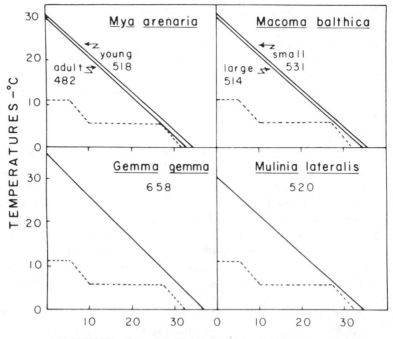

FIGURE 4.10   Temperature tolerance triangles for four bivalve mollusk species (solid lines) in relation to the state of Maryland's thermal discharge regulations for tidal waters (broken line). Numbers refer to the areas of the triangles in square degrees C. Reproduced with permission from Figure 3 of Kennedy and Mihursky (1971).

hazard (e.g., the impact agent), for assessment of the relation between the distribution of time-to-mortality and the hazard (Section 3.4.1).

Computer programs for analysis of data from toxicological studies are in both SAS (Service 1972, Barr et al 1976) and BMD (Dixon 1973). Buikema et al (1974) use the BMD program for probit analysis in a study of the sensitivity of a rotifer to heavy metals.

It is often desirable to use biological criterion variables other than mortality for assessment of sublethal effects of an environmental factor under controlled conditions (Section 3.7). For example, Whittle and Flood (1977) study the effects of a pulp mill effluent on rainbow trout, and in addition to determining the 96 hour LC_{50} also determine the effects on growth rate and on flavor of the flesh. They point out that in this case flavor is both the most sensitive sublethal response *and* the most critical economic response. Hummon (1974) and Hummon and Hummon (1975) use life table analysis (see Birch 1953, Southwood 1966, and Poole 1974)

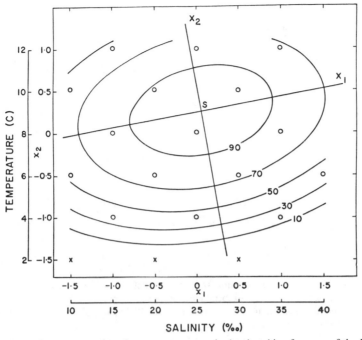

FIGURE 4.11   Response surface for percentage survival to hatching for eggs of the English sole *Parophrys vetulus* as a function of incubation temperature and salinity. The points indicate the original test conditions. Reproduced by permission of the Minister of Supply and Services Canada from Figure 12-24 of Alderdice (1972) after Figure 5 of Alderdice and Forrester (1968), *J. Fish. Res. Bd. Can.* **25**: 495–521.

to assess the effects of DDT on both mortality and reproduction in a gastrotrich. They argue that reproduction is usually more sensitive to environmental factors than is survival of the individual. The Leslie matrix analysis might well have been used instead of life table analysis (see above regarding transition matrices).

An effective format for presentation of the results of such studies is the temperature triangle technique of McErlean et al (1969), which is used by Kennedy and Mihursky (1971) to show upper temperature tolerances of estuarine bivalve mollusks (Figure 4.10). The dotted line in all cases represents the legal maximum temperature of a thermal discharge. Response, or trend, surfaces (Section 3.11) can be used effectively for display of results. Figure 4.11 from Alderdice (1972) shows percentage survival to hatching for eggs of the English sole as a function of salinity and temperature. Figure 4.12 combines information about lethal and sublethal effects from the same study. A perspective view of a response surface

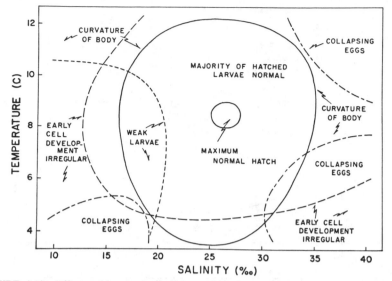

FIGURE 4.12 Effects of incubation salinity and temperature on egg development of the English sole. See Figure 4.11. Reproduced by permission of the Minister of Supply and Services Canada from Figure 12-25 of Alderdice (1972) after Figure 6 of Alderdice and Forrester (1968), *J. Fish. Res. Bd. Can.* **25**: 495–521.

describing the boundary of lethal conditions for lobsters for various combinations of temperature, salinity, and oxygen is presented by McLeese (1956) and is shown in Figure 4.13.

4.3 MAIN SEQUENCE 3: BASELINE OR MONITORING STUDY

In this main sequence we know that the impact has not yet occurred, but we do not know when or even if it will occur (columns 1 and 2 of Figure 3.2). When we know where the impact will occur, if it occurs, then both a control area and a potentially impacted area can be established (rows 3 and 4). Most often we will be monitoring at one (row 1) or more (row 2) locations within one area in order to detect impact if it occurs. If biological variables, such as species abundances, are used as the predictors of environmental impact effects, it is a biological monitoring study.

Baseline data are a prerequisite to any monitoring study, in that impact effects can only be detected as departures from the unimpacted state. The duration and complexity of the baseline study will depend on the extent to which there are natural trends and fluctuations in that state (see Sections 4.1, 4.2). If the biological response to the impact is unknown, either

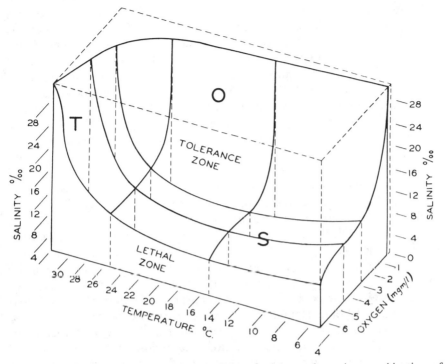

FIGURE 4.13 The boundary of lethal conditions for lobsters for various combinations of temperature, salinity, and oxygen. Reproduced by permission of the Minister of Supply and Services Canada from Figure 7 of McLeese (1956).

because the anticipated environmental change is unknown or it has never occurred before, only the H_0: "no biological change from the baseline state" is specified. Acceptance of H_0, or rejection of H_0 in favor of *any* H_A (represented by the biological community changing in any way), are the two possible outcomes. This is less satisfactory than posing a contrast between H_0: "baseline" and an H_A defined in terms of specified impact effects. It is the same choice that was discussed in Section 2.1.2, between H_A: "something is going on" and H_A: "this specified thing is going on." The latter kind of H_A cannot be defined on the basis of a baseline study alone. One or more impact studies conducted (by you or someone else) on the same system, or on systems that would be expected to respond in some way, must provide information about the biological response to the specified impact. The optimal design impact study presented as an example in Section 4.1 will be the basis of our "H_A defined" biological monitoring example in this section.

Some environmental biologists (e.g., Wilhm and Dorris 1968, Cairns 1974) have argued for the use of indices of community structure, such as species diversity indices, on the grounds that they are biological predictor variables whose response to deterioration of environmental quality is similar for many biological systems. See Section 3.5.2 for an extensive discussion of the use of diversity indices. Certainly, generality is a desirable attribute for a statistical model to be used in biological monitoring (Section 2.1.4). In practice, however, there is a tradeoff between the generality of the predictor and the useful information carried by the prediction. At one extreme foul odors, or dead animals lying all about, are pretty reliable indicators for any system of an environment gone bad. Usually we would like to know before that stage is reached, however. For most systems, decrease in some measure of species diversity is more likely to indicate an environmental change for the worse than a change for the better, but if one has gone to the effort to collect data necessary for calculation of a quantitative diversity index then there are probably better things that could be done with those data. Increase in abundance of indicator species has less generality, but can have more predictive power if the species are known to be characteristic of a particular environmental condition. Decrease or absence of particular species is more ambiguous (Sections 2.1.9, 3.4.1). Finally, at the other extreme are indices that are linear additive functions of biological predictor variables, derived from the results of impact studies such as the example in Section 4.1. They should not be generalized to systems other than the one for which they were derived, except with great caution, but they carry unambiguous information about rejection of a specified H_0 in favor of a specified H_A. The general subject of environmental indices was discussed in Section 3.5.1.

Returning to the example of Section 4.1, what criteria should be applied for future biological monitoring to detect impacts of the same kind? The best biological criterion will be the CVSI score (see Figure 4.1). Samples may be collected at any future time and a CVSI score (or average of scores) may be calculated from the species abundances $N_{j=1,3}$ for species $S_{j=1,3}$. If the estimated CVSI score exceeds some critical value, a warning of possible impact would go to the appropriate authorities, who would initiate corrective action or perhaps begin more extensive (and more costly) studies.

How is the critical CVSI score to be chosen? For this example one could simply look at Figure 4.1 and say that CVSI $= -1.6$ is a good dividing point. For a more rigorous approach, plot the 27 nonimpacted CVSI sample values as a cumulative percentage plot on arithmetic probability paper. Do the same for the 9 impacted sample values on the same plot. The points should form two lines that will be approximately straight if the distributions are normal and approximately parallel if the distributions have a common variance. (Canonical variates tend to be more normally distributed than were the variables from which they are calculated, so normality can generally be assumed.) Here we assume a common variance, which is in fact true given the simulation criteria. It is not necessary to do so, however. A pooled estimate of the common CVSI score variance is calculated from the usual ANOVA or t-test formula, and the mean is calculated for each group. The lines are then drawn on the arithmetic probability plot. Here they are superimposed on the CVSI-versus-CVEI scatter plot in Figure 4.1. It is now possible to determine the probability for any CVSI score of membership in either the impacted or nonimpacted group.

The choice of a critical value must be based on a balance of error risks that are weighted by the costs of making those errors, including social and aesthetic costs in addition to monetary ones. Say the only deciding factor is that the chance of a Type II error (deciding that there are no impact effects when in fact there are impact effects) must be no larger than 1 percent on each decision. Then the critical CVSI score is -2.17. Of course, at this critical value you will make the Type I error (deciding there are impact effects when in fact there are not) on 33 percent of your decisions, which will mean quite a few false alarms. If you monitor frequently and figure that marginal impact effects that are missed one time will probably be detected the next time when they are a little bit worse, you might decide that it is best to be conservative and reverse these error probabilities (CVSI $= -1.32$, α = chance of a Type I error = 1 percent, β = chance of Type II error = 33 percent). A third possibility would be to balance the two error risks at $\alpha = \beta = 8.5$ percent and choose CVSI $= -1.75$ as the critical value.

Sequential sampling designs (Sections 2.3.9, 3.8) could be applied to the biological monitoring situation. In the situation described above, for example, sampling could continue at each time until, say, H_0: "CVSI < -1.87" was accepted or rejected in favor of H_A: "CVSI > -1.63" with specified error risks. Alternatively, a hierarchical series of categories could be defined, as in Figure 2.12, and sampling would continue until one of them was entered. If CVSI < -1.87 were the category entered, then it could be assumed that all is well. If $-1.75 <$ CVSI < -1.63, then a "ready alert" might be announced, and entry into CVSI > -1.32 would trigger a full emergency alert.

What if H_A is not defined, and we wish to detect significant change of *any* kind from the baseline state? Here we wish to detect outliers (Section 3.9), and the statistical model is that of row 1 in Figure 3.3. The following example is based on the "before" data of Table 4.1, which consists of abundance estimates S_j for $j = 1.3$ species from three samples at each of six locations. The mean values of X_j for log-transformed abundances of species j at each location are

X_1	X_2	X_3
4.5	2.9	3.0
4.9	4.1	3.1
4.2	3.5	3.3
4.1	3.8	2.9
4.7	3.6	3.6
4.4	3.7	3.5

with means $\overline{X}_j$ taken over all six locations of

$$[4.47 \quad 3.60 \quad 3.23]$$

and variance-covariance matrix D of

$$\begin{bmatrix} 0.0907 & 0.0280 & 0.0213 \\ 0.0280 & 0.1600 & 0.0100 \\ 0.0213 & 0.0100 & 0.0787 \end{bmatrix}.$$

We want to test against H_0: "some *new* observation on log-transformed mean abundances $\hat{X}_j$, based on three replicate random samples collected at a randomly selected location, is a member of the trivariate normal baseline distribution corresponding to the data above." With p variables that test is

$$X^2 (p \text{ df}) = (\hat{X}_j - \overline{X}_j) D^{-1} (\hat{X}_j - \overline{X}_j)'.$$

If the new observation $\hat{X}_j$ is

$$[5.30 \quad 2.48 \quad 3.74],$$

then

$$X^2 (3 \text{ df}) = [0.83 \quad -1.12 \quad 0.51] \begin{bmatrix} 12.3600 & -1.9696 & -3.0950 \\ -1.9696 & 6.6139 & -0.3073 \\ -3.0950 & -0.3073 & 13.5830 \end{bmatrix} \begin{bmatrix} 0.83 \\ -1.12 \\ 0.51 \end{bmatrix}$$

$$= 21.7 \ (p < 0.01).$$

We therefore reject H_0 on the grounds that it is sufficiently improbable that an observation deviating this much from the mean values of the baseline distribution would occur by chance alone.

Suppose that we wish to reduce the p-dimensional data and the associated test to a two-dimensional graphical format. The two dimensions retaining the maximum information (= variation) in the data are those represented by the first two principal components $Y_{i=1,2}$ (Sections 3.4.3, 3.5.3, 3.9). A principal components analysis on these $n = 6 - \text{by} - p = 3$ data yields roots and vectors

λ_i (%)	$Y_i = f(X_j)$
0.1729 (52)	$Y_1 = 0.3584X_1 + 0.9165X_2 + 0.1783X_3$
0.0949 (29)	$Y_2 = 0.5916X_1 - 0.6561X_2 - 0.4686X_3$
0.0616 (19)	

The first two principal components retain 81 percent of the total information. The new variable Y_1 largely describes overall abundance variation, and Y_2 describes the variation in proportional abundance of the first species relative to the other two (see Section 3.5.3). Note that the λ_i, which are the variances on the new variables Y_i, add up to 0.3294 which is the sum of the original variances (in the diagonal of the D matrix). The covariances among the Y_i are of course zero. Therefore the $1 - \alpha$ probability ellipse in $Y_1 - Y_2$ space is

$$X^2_{1-\alpha} (2 \text{ df}) = [\hat{Y}_1 - \overline{Y}_1 \quad \hat{Y}_2 - \overline{Y}_2] \begin{bmatrix} \lambda_1 & 0 \\ 0 & \lambda_2 \end{bmatrix} \begin{bmatrix} \hat{Y}_1 - \overline{Y}_1 \\ \hat{Y}_2 - \overline{Y}_2 \end{bmatrix}$$

The Y_i means, $\overline{Y}_i$, are

$$\overline{Y}_1 = 0.3584\overline{X}_1 + 0.9164\overline{X}_2 + 0.1783\overline{X}_3 = 5.48$$

$$\overline{Y}_2 = 0.5916\overline{X}_1 - 0.6561\overline{X}_2 - 0.4686\overline{X}_3 = -1.23.$$

If we set $\alpha = 0.05$ and define $Y_i' = \hat{Y}_i - \overline{Y}_i$, then

$$5.991 = [Y_1' \quad Y_2'] \begin{bmatrix} 5.7837 & 0 \\ 0 & 10.5374 \end{bmatrix} \begin{bmatrix} Y_1' \\ Y_2' \end{bmatrix}$$

$$5.991 = 5.7837Y_1'^2 + 10.5374Y_2'^2$$

which is the equation for an ellipse with major axis proportional to the standard deviation $\sqrt{\lambda_1}$ on Y_1 and minor axis proportional to the standard deviation $\sqrt{\lambda_2}$ on Y_2. The ellipse is a function of deviations Y_i' of observed principal component scores $\hat{Y}_i$ from the baseline mean values $\overline{Y}_i$, and therefore is centered at $(Y_1', Y_2') = (0, 0)$. The $1 - \alpha = 0.99$ or any other probability ellipse would be calculated similarly, using the appropriate $X^2_{1-\alpha}$ value. The .95 and .99 probability ellipses are shown in Figure 4.14. The point indicated is for the new observation $\hat{X}_j = [5.30 \ 2.48 \ 3.74]$ from which is calculated $\hat{Y}_i = [4.84 \ -0.24]$ and $Y_i' = [-0.64 \ 0.99]$. The coordinates for the observation are therefore $(Y_1', Y_2') = (-0.64, 0.99)$. Since the point lies outside the .99 probability ellipse (Figure 4.14) we reject H_0, as before, at $p < 0.01$.

All calculations for the example above were done on a Texas Instruments SR-52 calculator. For large p or n a computer would be used, of course. For further discussion and presentation of examples for confidence ellipses and the calculation of classification probabilities see Jolicoeur and Mosimann (1960), Sokal and Rohlf (1969), Cooley and Lohnes (1971), and Davis (1973). Sokal and Rohlf present bivariate ellipse calculations in step-by-step cookbook form, rather than in matrix notation. Orloci (1972) provides an extensive review of "resemblance functions" in general, which may be measures of distance, probability, or information.

It should be remembered that the Type I error level as used in the monitoring examples of this section is for a single test. If tests, each at $\alpha = 0.05$, are made for observations at a series of five times, the probability of at least one significant result—assuming that the tests are independent and that H_0 is true throughout—is $1-(.95)^5 = 0.23$. For a 0.05 probability of no significant results on any of five independent tests, the α level for each test would have to be set at $\alpha = 1-(.95)^{1/5} = 0.01$. If monitoring will continue indefinitely, the best approach is to decide how many false alarms

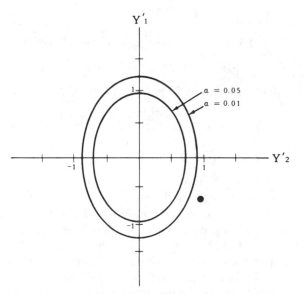

FIGURE 4.14 The .95 and .99 probability ellipses on the Y_1' and Y_2' axes for the "H_A undefined" example. The position of the new observation is shown.

you can afford (balanced against the number of missed impacts you can afford, of course) and then set α accordingly.

Finally, one can monitor levels of the impact agent directly, either in samples from the environment or in organisms. Use of uptake by or concentrations in organisms as biological variables was discussed in Section 3.7. The necessity for use of such variables if it is not known when and where the impact occurred is discussed in Section 4.5. Biologically active substances such as radioisotopes, pesticides, and heavy metals are often biologically monitored in this way. Examples are strontium 90 in reindeer antlers and clamshells where it replaces calcium, DDT taken up and biologically magnified as it passes up the food chain, and mercury which is methylated by bacterial action in lake and river sediments and then in this organic form is rapidly taken up and concentrated by organisms.

Large-scale monitoring, even worldwide, is now feasible with developments in aerial and satellite remote sensing, radar and sonar methods, and digital picture processing. A recent reference on this subject is edited by Schanda (1976).

4.4 MAIN SEQUENCE 4: IMPACT MUST BE INFERRED FROM SPATIAL PATTERN ALONE

Too often the environmental biologist is funded to study an impact after its effects have become a problem, and no before-impact data can be collected. In this case the impact effects must be demonstrated and described from spatial pattern. The options are those in columns 7 and 8 of Figure 3.2. Since areas differing in degree of impact cannot be chosen *a priori*, the first two rows are most relevant, and since spatial pattern must be the major source of information, the sampling design of row 2 and column 7 is most common.

The data matrix is usually of the form shown in the second row of Figure 3.3, with biological criterion variables from samples that have associated observations on environmental variables. Some replication of sampling at each location is always advisable (Sections 2.3.2, 2.3.3). The statistical analysis is likely to proceed in two stages: (1) reduction of the biological data to fewer variables that are efficient carriers of the information and (2) relating those biological variables to the environmental predictor variables in some explanatory manner. The biological data submatrix and the analysis models for the first stage correspond to the first row of Figure 3.3., with ordination and clustering methods being most appropriate. Ordination will produce a reduced number of continuous biological criterion variables, which would then be related to the environmental data submatrix by a statistical model of the form shown in the second row. Clustering methods will produce groups of sample locations with differing but internally homogeneous biological characteristics, and these "group variables" would then be related to the environmental data by a statistical model of the form shown in the third row. Biological species data could also be reduced to diversity indices but this is not recommended (see Sections 3.5.2 and 4.3 for discussion).

Much of the following material in this section is based on a presentation in a more condensed form by Green and Vascotto (in press). Both ordination and clustering methods have been discussed previously in relation to particular principles and problems (Sections 2.1.2, 2.3.9, 3.4.2, 3.4.3, 3.5.3, 3.9, 3.10, 3.11, 4.1, and 4.2). Here only an overview of these methods is attempted to guide the reader in choosing from a massive literature and a varied array of techniques. For an introductory treatment of ordination and clustering methods see Pielou (1969), Blackith and Reyment (1971), Marriott (1974), Poole (1974), and Clifford and Stephenson (1975). Williams and Lance (1968), Rohlf (1970), and Austin (1972) review ordination and clustering as they relate to each other and discuss the implications for choice of strategy. The review paper by Crovello (1970) provides an extensive bibliography. See Lefkovitch (1976) for demonstration of the close relationship between clustering and principal components or coordinates analysis.

An ordination transforms a data set containing n observations (samples) on p variables (say species abundances) into a reduced data set containing n observations on $k < p$ variables. The dimensionality is usually reduced from p to k in some manner that minimizes the loss of information caused by the reduction. Basic review papers on ordination methods are by Anderson (1971), Gauch and Whittaker (1972), and Orloci (1973b, 1974). Ordination is described in geometric terms by Orloci (1966) and Gower (1967a). Data standardization and transformation for ordination are reviewed by Orloci (1967) and Noy-Meir et al (1975).

Most commonly used ordination procedures are based on a model that assumes a linear additive relationship between the k new axes (factors, components, gradients, etc.) on the one hand and the p original variables (species abundances say) on the other. The hope is that each of the new axes describes a spatial pattern of species abundance determined by a dominant environmental factor. This is in contrast to the previously described uses of PCA, for example as an appropriate transformation method for conversion of data to quantitative form (Section 3.4.3), for the reduction of the dimensionality of the data (Sections 3.9, 3.11), or for deriving size and shape components (Section 3.5.3). However, when the samples cover a wide range of different environmental conditions and different species compositions related to those conditions, one should not expect the relationship between species abundances and an environmental factor to be linear or even monotonic (Gauch and Whittaker 1972). See Sections 2.3.9, 3.4.2, and 3.9 for previous mention of this problem. Many environmental biologists throw species abundance data into computer ordination programs, most commonly factor analysis (FA), forgetting that in their first ecology course they learned that species have optima on environmental gradients—unimodal distributions rather than linear ones (Figure 4.15). Misleading results are obtained when a linear model ordination (PCA, FA) is carried out using such data (Swan 1970, Noy-Meir and Austin 1970, Austin and Noy-Meir 1971, LaFrance 1972). When they are applied to species distributed along one environmental gradient, usually a curve in a *two*-dimensional ordination space results. This has been called the "horseshoe effect" (Kendall 1975, Fasham 1977). Noy-Meir (1974) concludes that "attempts to interpret each axis as an environmental factor are likely to produce anything from slightly distorted to wholly misleading results."

The simplest approach to nonlinear ordination is to apply linearizing transformations to the species abundance data. However, for relationships that are unimodal there is no *a priori* basis for choosing an appropriate

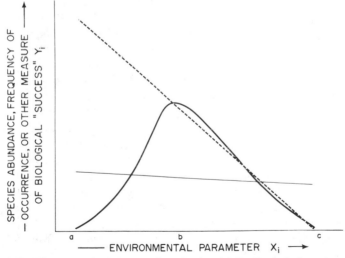

FIGURE 4.15 Diagrammatic representation of unimodal distribution of species success on an environmental gradient. If sampling covers the range a–c, then a linear model will show no relationship (thin solid line). If sampling covers the range b–c, then a significant relationship will be found, but one which is valid only over that range (broken line).

transformation. "Gaussian ordination" fits normal curves to species distributions on environmental variables (Gauch and Chase 1974, Gauch et al 1974, Ihm and Groenewoud 1975), but is limited to situations where there is only one dominant environmental gradient. Nonmetric multidimensional scaling (MDS) relaxes the linearity assumption but still assumes monotonicity (Section 3.4.2), which is too restrictive when wide ranges of environmental variables are involved and unimodal distributions of species on them are likely. For example, Fasham (1977) recently showed that nonmetric MDS does produce more interpretable results than, say, PCA but "It does not, however, manage to completely straighten out the horseshoes. . . ."

Some applications of ordination methods to analysis of spatial patterns in natural communities are for factor analysis (Dagnelie 1965, Bird 1970), for PCA and related methods (Allen 1971, Erman and Helm 1971, Blanc et al 1976, Sprules 1977), for PCA methods combined with clustering methods (Hughes 1971, Smith and Greene 1976, Reys 1976), and for PCA followed by discriminant analysis (James 1971). For computer programs useful in ordination see Section 3.10.

Clustering methods are appropriate when the hypothesis is that "the population of elements can most profitably be regarded as comprising an unknown number of partly-dissociated sub-populations" (Williams and Lance 1968). See also Section 3.9. The question of whether biological communities are continuous or discontinuous has long been debated [for reviews see Mills (1969) regarding marine benthic communities and Whittaker (1962) regarding terrestrial plant communities], but there is general agreement that successful application of clustering or ordination methods does not depend on the answer to this philosophical question (Lambert and Dale 1964, Anderson 1965, Goodall 1973b). Principles of clustering are reviewed by Williams (1971), and a classification of clustering strategies is presented and discussed. A thorough treatment of hierarchical methods, where groups are formed either by successive fusion of samples (agglomeratively) or by successive division into groups and subgroups (divisively), is given by Goodall (1973b). Recent books on cluster analysis methods in general are by Anderberg (1973) and Duran and Odell (1974). Gower (1967b) compares three popular cluster analysis methods, and the user's manual for the CLUSTAN package (Wishart 1975) discusses strategies for choosing a clustering procedure appropriate to the problem.

In addition to the general references given above, the following apply to clustering in special situations:

A mixture of data types—Lance and Williams (1967a, b, 1968).
Very large data sets—Crawford and Wishart 1967, 1968, Lefkovitch 1976.

Finding overlapping clusters—Day 1969, Wolfe 1970, Jardine and Sibson 1968, 1971.

Philosophy of testing in cluster analysis—Williams and Lance 1965, Sneath 1967.

Practice of testing in cluster analysis—Goodall 1966b, Wolfe 1970, Bottomley 1971, Maile 1972.

Comparing two or more cluster analyses—Rohlf 1974.

Clustering of sequences—Dale et al 1970.

Clustering to optimize prediction of an external criterion—Mac-Naughton-Smith 1963.

Clustering to group means for interpreting ANOVA results—Scott and Knott 1974.

Some examples of the application of clustering methods to the interpretation of environmentally controlled spatial pattern in biological communities are the following:

"Association analysis" (see Sections 3.4.1, 3.6, and 4.2) applied to terrestrial plant communities—Williams and Lambert 1959, 1960; applied to a marine intertidal community—Green and Hobson 1970.

Goodall's (1966b) "probability clustering" applied to marine phytoplankton—Legendre 1973.

An unweighted pair-group method applied to river benthic fauna affected by acid and fly-ash spills—Crossman et al 1974.

Identification of impacted and unimpacted lake benthic communities from a trellis diagram (Section 3.11) in relation to a pulpmill effluent —Sandilands 1977.

Computer programs for clustering are described in Section 3.10.

For methods relating the reduced biological data to the explanatory environmental data any model that assumes a linear relationship between species success (abundance, presence, frequency, diversity, or any other measure) and environmental variables is inappropriate. Such a relationship is assumed by simple linear or multiple regression of species on environmental variables, but it is also assumed by multivariate methods such as canonical correlation, PCA, or FA when both biotic and environmental variables are included in the same analysis. In the following example the first stage of the analysis, the efficient reduction of species abundance data, is accomplished by a cluster analysis. The resulting groups, or clusters, of samples are (by definition) characterized by relatively homogeneous species-assemblages that must be related in some manner to the environmental variables. The clusters can be plotted, sample by sample, on the map of the study area so that spatially contiguous clusters will stand out. However, in some studies, such as the example that follows, the spatial pattern of sample locations is not likely to reflect broad spatial patterns in the environmental variables. It would always be useful to quantitatively evaluate the separation, if any, of the species-assemblage groups on the environmental variables. A very simple approach is to carry out a one-way analysis of variance for each environmental variable in order to determine whether the groups, or clusters, of samples differ significantly from each other in their mean value on that variable (e.g., see Green and Hobson 1970).

Evaluation of significance of group separation and interpretation of such separation is best done by performing a multiple discriminant analysis (= canonical analysis) and then plotting the species-assemblage groups in the reduced discriminant space. Visually judging group separation on ecologically interpretable discriminant functions is in fact likely to be a more conservative approach than is use of the extremely powerful multivariate tests of significance (Green 1974). Discriminant analysis is illustrated by a simple example in Figure 4.16. There are two groups of samples, A and B, each presumably with a different species composition, and there are $p = 2$ environmental variables: X_1 = salinity and X_2 = temperature. The discriminant function is the linear function $b_1 X_1 + b_2 X_2 = 0$ with the b_j such that the separation of the groups is maximized. Here it is a stronger function of temperature than of salinity ($b_2 > b_1$) because temperature contributes more to the separation. Note that the assumed linear relationship is among the environmental variables rather than among the biotic variables or between the biotic variables and the environmental variables. In practice, logarithmic transformation of the environmental variables will often be appropriate (Section 2.3.9). The distribution of the biotic variables (the species-assemblage sample clusters) on the environmental variables is assumed to be roughly multivariate normal. However, if no formal tests of significance are made, then even this assumption is unimportant. Certainly the unimodality that usually characterizes such distributions will be adequate. Nonparametric approaches are available (Kendall 1966, Mantel 1970).

Regardless of the number of groups or the number of environmental variables the principle is the same, except that the possible number of discriminant functions is equal to the number of environmental variables or one less than the number of groups, whichever is smaller. For an introductory treatment of discriminant analysis see Cooley and Lohnes (1962, 1971), Marriott (1974), and Pimentel (1978). Several widely available computer packages can be used for these analyses, including BMD, SAS, and SPSS (see Section 3.10). It is possible to carry out a stepwise discriminant analysis, analogous to a stepwise multiple regression analysis,

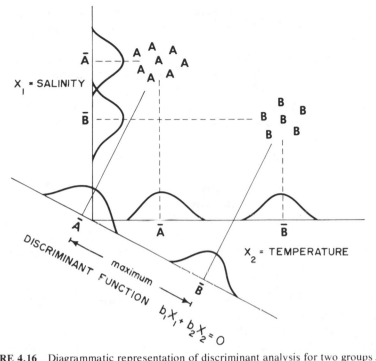

FIGURE 4.16 Diagrammatic representation of discriminant analysis for two groups A and B and two environmental variables X_1 and X_2. The discriminant function is a new axis that maximizes the separation of the groups. Reproduced with permission from Figure 1 of Green and Vascotto (in press).

but this is not recommended. It is generally assumed that the discriminant functions (DF's) themselves have ecological interpretations, as in the example that follows, and these interpretations are best made from the standardized DF coefficients $b_{j=1,p}$ on *all* of the p environmental variables. If one could assume that the original environmental variables were themselves the independent environmental factors controlling species distributions, then a stepwise analysis might be appropriate (see Section 3.6). This would rarely be the case. Another possibility is to seek rotated DF axis solutions, as is common in factor analysis methodology. For the use described here the unrotated solution is generally the most interpretable one.

The biological data are taken from Patalas (1971) and describe zooplankton species abundance for 27 species in 34 lakes in northwestern Ontario. Environmental data taken from Patalas (1971) and Armstrong and Schindler (1971) consist of values for 11 morphometric and chemical variables (area, maximum depth, conductivity, total dissolved solids, Ca, Mg, Na, K, Fe, N, and P) for each of the 34 lakes. Each lake, therefore, is a sample.

The result of a hierarchical cluster analysis is shown in Figure 4.17. The similarity coefficient used is the measure of mutual information proposed by Orloci (1968). Programs in the BASIC language for carrying out the analysis are given in Orloci (1975a). Four distinct groups of lakes with differing zooplankton species assemblages (Table 4.6) appear to exist. The discriminant analysis will therefore be an assessment of separation of four groups in an 11-dimensional environmental space.

In Table 4.6 the species are listed in order of the magnitude of their contribution to faunal group definition, as measured by the F-statistic (the ratio of the among-groups to the within-groups variance of log-abundance). These are not used here as tests of significance, but rather as indices of the degree to which the faunal groups are defined by each species (see Section 3.5.4). For each species the .95 confidence limits are based on the error mean square (the denominator from the F-statistic calculation) and are presented as an aid in assessing which group or groups are defined by that species. The best key to these groups based on indicator species is presented in Table 4.7. See Green and Vascotto (in press) for discussion of the biological associations.

Figure 4.18 shows the separation of the groups of lakes on the first two discriminant functions, which account for 54 and 39 percent, respectively, of the among-group variance (Table 4.8). Multivariate analysis of variance

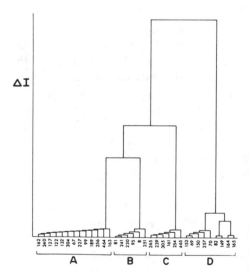

FIGURE 4.17 Hierarchical cluster analysis of 34 lakes based on abundances of 27 zooplankton species. The vertical axis represents the information gain on fusion (ΔI), the higher the level of fusion, the more dissimilar are the lakes in species composition. On the horizontal axis are lake code numbers and group code letters. Reproduced with permission from Figure 1 of Green and Vascotto (in press).

Table 4.6 Comparison of faunal groups on biotic variables

Species	F (pr.) (3, 30 df)	Geometric Mean Number of Individuals per Square Centimetre (.95 Confidence Limit)			
		Group A	Group B	Group C	Group D
Diaptomus oregonensis	41.0 (<.01)	0.00 (0–0.01)	0.01 (0–0.05)	0.00 (0–0.02)	1.6 (0.68–3.8)
Diaptomus minutus	35.0 (<.01)	7.0 (2.7–18)	0.04 (0–0.25)	13.0 (2.5–68)	0.02 (0–0.09)
Cyclops bicuspidatus	10.0 (<.01)	1.3 (0.27–6.1)	0.03 (0–0.57)	21.0 (1.5–313)	0.03 (0–0.27)
Epischura lacustris	9.1 (<.01)	0.14 (0.06–0.33)	0.00 (0–0.03)	0.08 (0.01–0.36)	0.00 (0–0.02)
Diaptomus leptopus	7.5 (<.01)	0.00 (0–0.01)	0.14 (0.03–0.53)	0.00 (0–0.03)	0.00 (0–0.02)
Ceriodaphnia lacustris	4.3 (<.05)	0.00 (0–0.01)	0.00 (0–0.01)	0.00 (0–0.01)	0.02 (0.01–0.05)
Mesocyclops edax	4.3 (<.05)	0.37 (0.09–1.5)	0.03 (0–0.42)	0.08 (0–0.96)	1.32 (0.31–10)
Senecella calanoides	3.7 (<.05)	0.00 (0–0.01)	0.00 (0–0.01)	0.02 (0.01–0.05)	0.00 (0–0.01)
Diaphanosoma brachyurum	3.1 (<.05)	0.02 (0–0.07)	0.17 (0.02–1.1)	0.00 (0–0.05)	0.01 (0–0.06)
Bosmina longirostris	3.1 (<.05)	0.11 (0.03–0.38)	0.23 (0.02–1.8)	1.4 (0.17–11)	0.85 (0.18–3.8)

Table 4.7   Key to zooplankton species-assemblage groups $A-D$, based on best indicator species

1.a	*Diaptomus oregonensis* density >0.4 individuals/cm^2	*Group D*
1.b	*D. oregonensis* density <0.4 individuals/cm^2. Go to 2	
2.a	Relatively depauperate lakes (<15 individuals/cm^2), with *Diaptomus minutus* <0.8 individuals/cm^2 and *D. leptopus* >0.03 individuals/cm^2	*Group B*
2.b	Go to 3 (*D. minutus* dominant)	
3.a	*Cyclops bicuspidatus* density >12 individuals/cm^2	*Group C*
3.b	*C. bicuspidatus* density <12 individuals/cm^2	*Group A*

and covariance was used to test whether lake area and depth alone provide significant separation of the species-assemblage groups (yes, $p < 0.01$) and whether the nine water chemistry variables provide significant separation *in addition to* that provided by area and depth (no, $p > 0.05$). We therefore proceed with interpretation in terms of the variables area and depth only. Groups D, A, and C are separated on DF I and form a series of increasing lake size, with lakes in all three groups exhibiting basin morphometries that allow good wind-generated circulation. Group B is separated from the other groups on DF II and differs from them in consisting of lakes with very small surface area (on the average, less than a third of that for group D lakes) but relatively great depth (on the average, more than twice that for group D lakes). These are lakes with poor water circulation.

A recent paper by Sprules (1977) provides an example of a different multivariate statistical approach to the Patalas data, although comparisons should be made with caution because the lakes used are not exactly the same. His approach is to perform a principal components analysis on the zooplankton species abundance data and then calculate rank correlations between the first two principal components (PC's) and six environmental variables. Similar analysis methodologies are used by Barkham and Norris (1970) and Blanc et al (1972) for studies of terrestrial vegetation and marine phytoplankton, respectively. Only the first PC appears to be at all meaningful and it is basically a separator of large, deep, and clear lakes from small, shallow, and more turbid lakes. In other words, it is roughly the equivalent of our DF I. The second PC appears to be the nonlinear component of this trend through the data. The horseshoe effect is evident. There is no descriptor of lake basin shape, separating the poor circulation (our group B) lakes.

In this example the initial information is in the abundances of 27 zooplankton species in 34 lakes that are characterized by 11 environmental

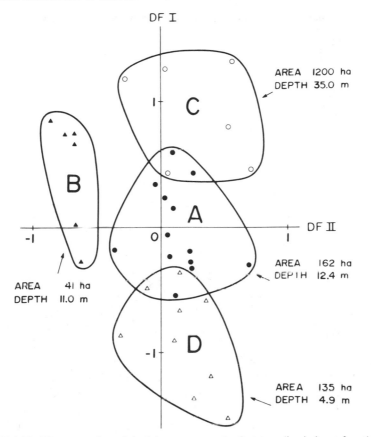

FIGURE 4.18 The separation of the lake groups on the first two discriminant functions of the 11 environmental variables. Mean values for area and maximum depth are given for each group.

variables. The cluster analysis—discriminant analysis results show that lakes clustered into groups with similar species-compositions occupy different positions in the environmental space. The original 11-dimensional environmental space can be reduced to 2 dimensions that represent 2 ecologically interpretable factors providing sufficient separation of the species-assemblage groups, and these 2 factors depend on only 2 of the 11 environmental variables. Formal tests show significance as well, but most important is the fact that realistic statistical models produce visual display which is in itself the convincing evidence of environmental control of biotic spatial patterns.

Two cautionary remarks should be made. First, the multiple discriminant analysis is particularly susceptible to rounding error and calculations

Table 4.8   Discriminant analysis of lake zooplankton data

Discriminant function	I	II	III
Percent of separation	54	39	7
Cumulative percent of separation	54	93	100
Variables and standardized discriminant function coefficients			
Area	−0.15	0.35	
Depth	0.59	−0.08	
Conductivity	0.30	−0.42	
Total dissolved solids	0.30	−0.26	
Ca	−0.48	0.68	
Mg	−0.11	−0.15	
Na	−0.06	0.04	
K	0.38	−0.25	
Fe	−0.04	0.08	
N	−0.07	0.16	
P	−0.20	−0.21	

in double precision for at least the eigenvalue-eigenvector routines are advisable. When a simplified (three environmental variables) version of this analysis was run on a Wang 462-1 calculator, which stores and calculates with 13 digits, it was found that the roots (eigenvalues) differed in the third significant digit and some vector (eigenvector) elements differed in the second significant digit from IBM 360 double precision results. Second, remember that in the absence of a before-impact temporal control conclusions based on main sequence 4 analyses always depend on the unverifiable assumption that the spatial pattern is impact related and did not exist before the impact. Spatial pattern analysis results, especially from transects across environmental gradients, can be misleading because of confounded variables (Section 2.1.9).

In the example above the samples are lakes distributed over a large area with complex geological variation and therefore their map positions are unlikely to reveal spatial pattern. For an example of cluster analysis followed by discriminant analysis for samples from a grid superimposed over broad spatial patterns, using the simulated data of Section 4.2, see Green (1977).

The second analysis stage—relating the reduced biological criterion variables to the environmental predictor variables—could be done in other ways. Cluster analysis could be performed on the environmental variables too, as done by Sandilands (1977) in relation to pulp mill effluent effects, and then the two cluster analysis results could be related by methods such as those described by Rohlf (1974). Two-way contingency table analysis is one possibility. Where the first analysis stage is an ordination the PC, FA, etc., scores are often related to the environmental variables by various linear or nonparametric correlation-type analyses (see above). Since there may be nonmonotonicity as well as nonlinearity, the first step should be to examine scatter plots of biological against environmental variable values. Obvious relationships could stand on their own without additional statistics or a curve-fitting method such as polynomial regression could be used. See Section 3.7 and Figure 3.16 regarding the successful application of a multivariate linear model (canonical correlation analysis) to relationships between morphological and environmental variable sets. The success of a linear model in such an analysis will depend on the environmental range and the type of community. For example, if samples are on a transect across an estuary and the species are marine at one end and are gradually replaced by freshwater species as one moves up the transect, a linear biological-environmental relationships model is likely to work (with transformations of variables). If, however, there are true estuarine species with modal success values (abundance, presence, size, etc.) at intermediate salinities, the model will fail.

4.5   MAIN SEQUENCE 5: "WHEN AND WHERE" IS THE QUESTION

The best possible environmental study design is described in Section 4.1, and the worst possible situation is described here. Again, as in Section 4.4, the options are those of columns 7 and 8 of Figure 3.2, but here without knowledge of when or where the impact first occurred. Determining when and where, and the extent to which the supposed impact is of natural rather than of human origin, must be the objectives.

In this situation, the strategy must be to determine whether the environmental impact in question existed at times before hypothesized human causes were operating and/or whether it now exists in places unaffected by human activity that might be the cause. The mercury pollution problem is a classic example of this situation, and is reviewed by Katz (1972). There is no doubt that some aquatic systems have extremely elevated levels in sediment, water, and tissues of organisms and are obviously influenced by industrial or agricultural sources of mercury. However, mercury compounds occur naturally in rocks (Evans et al 1972) and some lakes uninfluenced by modern human society contain fish with unsafe mercury levels. The use of artifacts as time machines for a situation such as this was mentioned in Section 3.7. Miller (1972) determined mercury concentrations in museum specimens of tuna and swordfish dating back 25 and 62 to 93 years. He found that levels for different ages fell into the same range and concluded that the perception of a mercury problem had much to do with improved methodology and increased awareness. However, Evans et al 1972) found that preserved fish from both Lake Erie and Lake St. Clair had higher mercury levels if they were collected in the period 1965–1970 than fish collected in 1945 or earlier. Weiss (1971) examined mercury levels in a Greenland ice sheet and concludes that input by man has resulted in an increase over the last several decades. In summary, it seems that natural mercury levels can be and in some cases always have been higher than those currently defined as safe, but recent human impact has increased the levels generally as well as in particular places.

A similar problem, described by Hickey and Anderson (1968), is that of determining whether certain raptorial bird species have declined because of the effects of chlorinated hydrocarbon pesticide residues on reproduction. They examined 1729 blown eggs from 39 museum and private collections and compared eggs of control species (those that have not been declining any faster than habitat destruction would explain) with eggs of declining raptorial species. The conclusion was that a reduction of shell thickness of at least 19 percent accompanied catastrophic declines of three raptorial species and that the onset of the declines began one year after the start of general usage of chlorinated hydrocarbons.

There is great potential for studies of this kind. With new techniques such as those described by Rhoads and Panella (1970) it is possible to use fossil shells to determine growth over annual, lunar monthly, daily, and even tidal cycles for organisms that were living 500 million years ago (Panella et al 1968). Runcorn (1966) describes similar use of corals. For terrestrial environments Haugen (1967) describes relationships between tree rings and climatic variation between 1650 and the present.The use of such growth rates to assess biological effects of past environmental conditions would seem to have great promise.

References

The numbers above each reference describe the contents in relation to the purpose for which it was cited:

1 Is a review paper or basic reference book.
2 Explains or provides calculations for a method.
3 Cites, uses, or lists a computer program.
4 Is related to a freshwater environment.
5 Is related to a marine or estuarine environment.
6 Is related to a terrestrial environment.
7 Describes or uses a multivariate method.
8 Involves diversity or related concepts.
9 Uses or discusses simulation methods.

The sections of the book in which the reference is cited are indicated in brackets following each reference.

1 5 7
Alderdice, D. F. 1972. Factor combinations. Responses of marine poikilotherms to environmental factors acting in concert. In *Marine ecology*, Vol. 1, *Environmental factors*, Part 3, O. Kinne (Ed.), pp. 1659–1722. Wiley, New York [4.2]
4 5
Allee, W. C., A. E. Emerson, O. Park, T. Park, and K. P. Schmidt. 1949. *Principles of animal ecology*. Saunders, New York. [3.5.2]
2 6 7
Allen, T. F. 1971. Multivariate approaches to ecology of algae on terrestrial rock surfaces in North Wales. *J. Ecol.* **59**: 803–826. [3.4.1, 3.4.5, 4.4]
1
American Mathematical Society. 1962. A manual for authors of mathematical papers. *Bull. Am. Math. Soc.* **68**: 1–16. [2.1.8]

1 2 3 7

Anderberg, M. R. 1973. *Cluster analysis for applications*. Academic Press, New York. [3.4.1, 3.10, 4.4]

1 7

Anderson, A. J. 1971. Ordination methods in ecology. *J. Ecol.* **59**: 713–726. [3.4.2, 3.9, 4.4]

1 7

Anderson, D. J. 1965. Classification and ordination in vegetation science: Controversy over a nonexistent problem. *J. Ecol.* **53**: 521–576. [4.4]

7

Anderson, R. S. 1971. Crustacean plankton of 146 Alpine and Subalpine lakes and ponds in Western Canada. *J. Fish. Res. Bd. Can.* **28**: 311–321. [3.11]

1 7

Anderson, T. W. 1963. The use of factor analysis in the statistical analysis of multiple time series. *Psychometrika* **28**: 1–25. [4.2]

1 2

Anderson, T. W. 1971. *The statistical analysis of time series*. Wiley, New York. [4.2]

1 2 7

Andrews, D. F., R. Gnanadesikan, and J. L. Warner. 1971. Transformations of multivariate data. *Biometrics* **27**: 825–840. [2.3.9]

2

Anscombe, F. J. 1948. The transformation of Poisson, binomial and negative-binomial data. *Biometrika* **35**: 246–254. [2.3.9]

1 2

Anscombe, F. J. 1950. Sampling theory of the negative binomial and logarithmic series distributions. *Biometrika* **37**: 358–382. [2.3.9]

3 7

APL/360 user's manual. 1969. IBM, White Plains, New York. [3.4.3, 3.6, 3.10, 3.11]

4 7 8

Archibald, R. E. M. 1972. Diversity in some South African diatom associations and its relation to water quality. *Water Res.* **6**: 1229–1238. [3.5.2]

4

Armstrong, F. A. J., and D. W. Schindler. 1971. Preliminary chemical characterization of waters in the Experimental Lakes Area, northwestern Ontario. *J. Fish. Res. Bd. Can.* **28**: 171–187. [4.4]

2

Arnason, A. N. 1973. The estimation of population size, migration rates and survival in a stratified population. *Res. Popul. Ecol.* **15**: 1–8. [3.7]

3

Arnason, A. N., and L. Baniuk. 1977. *User's manual—POPAN 2: A data maintenance and analysis system for recapture data*. Release 2. Charles Babbage Research Centre, University of Manitoba. [3.7, 3.10]

3 7

Atchley, W. R. 1971. A comparative study of the cause and significance of morphological variation in adults and pupae of *Culicoides*: A factor analysis and multiple regression study. *Evolution* **25**: 563–583. [3.7]

1 9

Atchley, W. R., C. T. Gaskins, and D. Anderson. 1976. Statistical properties of ratios. I. Empirical results. *Syst. Zool.* **25**: 137–148. [3.5.3]

6 7 8

Auclair, A. N. 1971. Diversity relations of upland forests in the western Great Lakes area. *Am. Nat.* **105**: 499–528 [3.5.2]

1 3 7

Austin, M. P. 1972. Models and analysis of descriptive vegetation data. In *Mathematical models in ecology*, J. N. R. Jeffers (Ed.) pp. 61–86. Blackwell, Oxford. [4.4]

1 6 7

Austin, M. P., and P. Greig-Smith. 1968. The application of quantitative methods to vegetation survey. II. Some methodological problems of data from rain forest. *J. Ecol.* **56**: 827–844. [2.3.9, 3.6]

7 9

Austin, M. P., and I. Noy-Meir. 1971. The problem of non-linearity in ordination experiments with two-gradient models. *J. Ecol.* **59**: 763–773. [4.4]

2

Bailey, N. T. J. 1951. On estimating the size of mobile populations from recapture data. *Biometrika* **38**: 293–306. [3.7]

2 6

Banerjee, B. 1969. A mathematical model on sampling diplopods using pitfall traps. *Oecologia* **4**: 102–105. [3.7]

6 7

Barkham, J. P., and J. M. Norris. 1970. Multivariate procedures in an investigation of vegetation and soil relations of two beech woodlands, Cotswold Hills, England. *Ecology* **51**: 630–639. [4.4]

1 2

Barnes, H. 1952. The use of transformations in marine biological statistics. *J. Cons.* **18**: 61–71. [2.3.9]

5

Barnett, P. R. O. 1971. Some changes in intertidal sand communities due to thermal pollution. *Proc. Roy. Soc. Lond. B* **177**: 353–364. [3.2]

3 7 9

Barr, A. J., J. H. Goodnight, J. P. Sall, and J. T. Helwig. 1976. *A user's guide to SAS 76.* SAS Institute Inc. Raleigh, NC. [2.1.7, 3.4.2, 3.4.3, 3.6, 3.10, 3.11, 4.1, 4.2]

1 9

Barrett, J. P., and L. Goldsmith. 1976. When is *n* sufficiently large? *Am. Stat.* **30**: 67–70. [2.3.8]

1 2

Bartlett, M. S. 1947. The use of transformations. *Biometrics* **3**: 39–52. [2.3.9]

2

Bartlett, M. S. 1954. A note on the multiplying factors for various chi-square approximations. *J. Roy. Stat. Soc. B* **16**: 290–298. [3.9]

2 8

Barton, D. E., and F. N. David. 1959. The dispersion of a number of species. *J. Roy. Stat. Soc. B* **21**: 190–194. [3.5.2]

1 2

Batschelet, E. 1976. *Introduction to mathematics for life sciences*. Springer-Verlag, New York [1.2, 3.11, 4.2]

4

Bedford, J. W., E. W. Roelofs, and M. J. Zabik. 1968. The freshwater mussel as a biological monitor of pesticide concentration in a lotic environment. *Limnol. Oceanogr.* 13: 118–126. [3.7]

2

Berkson, J. B. 1944. Application of the logistic function to bioassay. *J. Am. Stat. Assoc.* 39: 357–365. [4.2]

2

Berkson, J. B. 1946. Approximation of chi-square by probits and logits. *J. Am. Stat. Assoc.* 41: 70–74. [4.2]

2

Berkson, J. B. 1951. Why I prefer logits to probits. *Biometrics* 7: 327–339. [4.2]

2 4 5

Beverton, R. J. H., and S. J. Holt. 1957. On the dynamics of exploited fish populations. *U. K. Min. Agr. Fish., Fish. Invest. Ser.* 2, 19. [3.7]

2 7

Bhargava, R. P. 1972. A test for equality of means of multivariate normal distributions when covariance matrices are unequal. *Calcutta Stat. Assoc. Bull.* 20: 153–156. [4.1]

2 6

Birch, L. C. 1953. Experimental background to the distribution and abundance of insects. I. The influence of temperature, moisture and food on the innate capacity for increase of three grain beetles. *Ecology* 34: 698–711. [4.2]

5 7

Bird, S. O. 1970. Shallow marine and estuarine benthic molluscan communities from area of Beaufort, North Carolina, *Am. Assoc. Pet. Geol.* 54: 1651–1676. [2.1.9, 4.4]

4 7

Bishop, M. J., and H. DeGaris. 1976. A note on population densities of Mollusca in the River Great Ouse at Ely, Cambridgeshire. *Hydrobiologia* 48: 195–197. [2.3.9]

1 2 3 7

Blackith, R. E., and R. A. Reyment. 1971. *Multivariate morphometrics*. Academic Press, London. [3.4.3, 3.5.3, 4.4]

5 7

Blanc, F., P. Chardy, A. Laurec, and J. P. Reys. 1976. Choix des métriques qualitative en analyse d'inertie. Implications en écologie marine benthique. *Mar. Biol.* 35: 49–67. [4.4]

5 7

Blanc, F., M. Leveau, M. C. Bonin, and A. Laurec. 1972. Écologie d'un milieu eutrophique: traitement mathématique des données. *Mar. Biol.* 14: 120–129. [4.4]

2

Bliss, C. J. 1953. Fitting the negative binomial distribution to biological data. *Biometrics* 9: 176–200. [2.3.9]

3 5 7

Bloom, S., J. Simon, and V. Hunter. 1972. Animal sediment relations and community analysis of a Florida estuary. *Mar. Biol.* 13: 43–56. [2.3.6, 3.11]

5 7

Bodiou, J. Y., and P. Chardy. 1973. Principal component analysis of the annual cycle of an harpacticoid copepod assemblage from the infralittoral fine sands of Banyuls-sur-Mer. *Mar. Biol.* **20**: 27–34. [2.3.9]

1 3

Boillot, M., and L. W. Horn. 1976. *BASIC.* West Publ. Co., New York. [3.10]

2 7 9

Borucki, W. J., D. H. Card, and G. C. Lyle. 1975. A method of using cluster analysis to study statistical dependence in multivariate data. *IEEE Trans. Comput.* **24**: 1183–1191. [3.9]

1 7

Bottomley, J. 1971. Some statistical problems arising from the use of the information statistic in numerical classification. *J. Ecol.* **59**: 339–342. [4.4]

3

Bower, C. P., W. L. Padia, and G. V. Glass. 1974. *TMS: Two Fortran IV programs for the analysis of time-series experiments.* Laboratory of Educational Research, University of Colorado, Boulder, Colo. [3.10, 4.2]

8

Bowman, K. O., K. Hutcheson, E. P. Odum, and L. R. Shenton. 1971. Comment on the distribution of indices of diversity. In *Statistical ecology*, Vol. 3, G. P. Patil, E. C. Pielou, and W. E. Waters (Eds.). Pennsylvania State University Press, University Park, Pa. 315–359. [3.5.2]

2

Box, G. E. P., and D. R. Cox. 1964. An analysis of transformations. *J. Roy. Stat. Soc.* B. **26**: 211–252. [2.3.9]

2 6 7

Bradfield, G. E., and L. Orloci 1975. Classification of vegetation data from an open beach environment in southwestern Ontario: Cluster analysis followed by generalized distance assignment. *Can. J. Bot* **53**: 495–502. [3.9]

2 4

Brett, J. R., 1956. Principles of thermal requirements of fish. *Q. Rev. Biol.* **31**: 75–87. [4.2]

5 7

Briand, F. J. P. 1975. Effects of power-plant cooling systems on marine phytoplankton. *Mar. Biol.* **33**: 135–146. [3.5.2]

8

Brillouin, L. 1962. *Science and information theory*, 2nd ed. Academic Press, New York. [3.5.2]

1 2 4 5 6 7 8

British Columbia Department of Lands, Forests and Water Resources (Ecology Section, Pollution Control Branch). 1974. A guide to some biological sampling methods. *Inf. Bull.* Victoria, B. C. [2.1.6, 3.5.2, 3.7]

2

Brown, G. H., and N. J. Fisher. 1972. Subsampling a mixture of sampled material. *Technometrics* **14**(3): 663–668. [2.3.2]

1 2 3

Brownie, C., D. R. Anderson, K. P. Burnham, and D. S. Robson, 1978. *Statistical inference from band recovery data—A handbook*. U. S. Fisheries and Wildlife Service, Resource Publ. No. 131, Washington, D.C. [3.7]

2 7

Bryant, E. H., and R. R. Sokal. 1967. The fate of immature housefly populations at high and low densities. *Res. Popul. Ecol.* 9: 19–44. [3.11]

2 3 4

Buikema, A. L., Jr., J. Cairns, Jr., and G. W. Sullivan. 1974. Evaluation of *Philodina acuticornis* (Rotifera) as a bioassay organism for heavy metals. *Water Resour. Bull.* 10: 648–661. [4.2]

2 6

Bulmer, M. G. 1974. A statistical analysis of the ten-year cycle in Canada. *J. Anim. Ecol.* 43: 701–718. [4.2]

5 7

Buzas, M. A. 1967. An application of canonical analysis as a method for comparing faunal areas. *J. Anim. Ecol.* 36: 563–577. [2.3.9]

5 7

Buzas, M. A. 1970. Spatial homogeneity: statistical analyses of unispecies and multispecies populations of Foraminifera. *Ecology* 51: 874–879. [2.3.8]

1 8

Buzas, M. A. 1972. Patterns of species diversity and their explanation. *Taxon* 21: 275–286. [3.4.1, 3.5.2]

1 4 7 8

Cairns, J., Jr. 1974. Indicator species vs. the concept of community structure as an index of pollution. *Water Resour. Bull.* 10: 338–347 [2.1.9, 4.3]

2 3 4 7 8

Cairns, J., Jr. and K. L. Dickson. 1971. A simple method for the biological assessment of the effects of waste discharges on aquatic bottom-dwelling organisms. *J. Water Pollut. Control Fed.* 43: 755–772. [1.1, 2.1.5, 3.6, 3.7, 3.10]

4

Cairns, J., Jr., K. L. Dickson, and G. F. Westlake, (Eds.). 1977. *Biological monitoring of water and effluent quality*. American Society for Testing and Materials, Philadelphia, Pa. [3.7]

1 4 7 8

Cairns, J., Jr., G. R. Lanza, and B. C. Parker. 1972. Pollution related structural and functional changes in aquatic communities with emphasis on freshwater algae and protozoa. *Proc. Acad. Nat. Sci. Phila.* 124: 79–127. [3.5.2]

1 4 7 8

Cairns, J., Jr., J. S. Crossman, K. L. Dickson, and E. E. Herricks. 1971. The recovery of damaged streams. *ASB Bull.* 18: 79–106. [3.5.2, 3.8]

4

Cairns, J., Jr., J. W. Hall, E. L. Morgan, R. E. Sparks, W. T. Waller, and G. F. Westlake. 1974. The development of an automated biological monitoring system for water quality management. In *Trace substances in environmental health—VII*, D. D. Hemphill, (Ed.), pp. 35–40., University of Missouri, Columbia, Mo. [3.7]

2 3 4 8

Carle, F. L. 1976. An evaluation of the removal method for estimating benthic populations and diversity. M. S. thesis. Virginia Polytechnic Institute and State University, Blacksburg, Va. [2.3.6, 3.7, 3.10]

2 3 9

Capra, J. R., and R. S. Elster. 1971. Notes on generating multivariate data with desired means, variances, and covariances. *Educ. Psych. Meas.* **31**: 749–752. [2.3.9, 3.10, 4.1]

1 2 3 7

Carmichael, J. W., and P. H. A. Sneath. 1969. Taxonometric maps. *Syst. Zool.* **18** : 402–415. **[3.11]**

2

Carothers, A. D. 1971. An examination and extension of Leslie's test of equal catchability. *Biometrics* **27**: 615–630. [3.7]

2

Cassie, R. M. 1954. Some uses of probability paper in the analysis of size-frequency distributions. *Austr. J. Mar. Freshwater Res.* **5**: 513–522. [3.5.3, 3.7]

12

Cassie, R. M. 1962. Frequency distribution models in the ecology of plankton and other organisms. *J. Anim. Ecol.* **31**: 65–92. [2.1.7, 2.3.9]

5 7

Cassie, R. M. 1972. Fauna and sediments of an intertidal mudflat: An alternative multivariate approach. *J. Exp. Mar. Biol. Ecol.* **9**: 55–64. [3.6, 3.9]

5 7

Cassie, R. M., and A. D. Michael. 1968. Fauna and sediments of an intertidal mud-flat: A multivariate analysis. *J. Exp. Biol. Ecol.* **2**: 1–23. [2.3.9]

1 2 7

Cattell, R. B. 1965a. Factor analysis: An introduction to essentials. I. The purpose and underlying models. *Biometrics* **21**: 190–215. [3.9]

1 2 7

Cattell, R. B. 1965b. Factor analysis: An introduction to essentials. II. The role of factor analysis in research. *Biometrics* **21**: 405–435. [3.9]

2 7 9

Chan, L. S. 1972. Treatment of missing values in discriminant analysis. I. Sampling experiment. *J. Am. Stat. Assoc.* **67**: 473–477. [3.5.3]

2

Chapman, D. G., and C. O. Junge, Jr. 1956. The estimation of the size of a stratified animal population. *Ann. Math. Stat.* **27**: 375–389. [3.7]

2

Chapman, D. G., and G. I. Murphy. 1965. Estimates of mortality and recruitment from a single-tagging experiment. *Biometrics* **21**: 921–935. [3.7]

2

Chatfield, C. 1969. On estimating parameters of logarithmic series and negative binomial distributions. *Biometrika* **56**: 411–414. [2.3.9]

2 5

Clark, G. R., II. 1968. Mollusc shell: daily growth lines. *Science* **161**: 800–802. [3.7]

6

Cleveland, W. S. 1976. Photochemical air pollution: Transport from the New York City area into Connecticut and Massachusetts. *Science* 191: 179–181. [3.8]

2

Cleveland, W. S., and B. Kleiner. 1975. A graphical technique for enhancing scatter plots with moving statistics. *Technometrics* 17: 447–454. [2.1.7, 3.11]

1 2 7 8

Clifford, H. T., and W. Stephenson. 1975. *An introduction to numerical classification.* Academic Press, San Francisco. [3.4.1, 3.4.3, 3.11, 4.4]

1

Cochran, W. G. 1947. Some consequences when the assumptions for the analysis of variance are not satisfied. *Biometrics* 3: 22–38. [2.1.7, 2.3.9, 3.9]

2

Cochran, W. G. 1950. The comparison of percentages in matched samples. *Biometrika* 37: 256–266. [3.4.1, 4.2]

1 2

Cochran, W. G. 1957. Analysis of covariance: Its nature and uses. *Biometrics* 13: 261–281. [3.5.3]

1 2

Cochran, W. G. 1963. *Sampling techniques,* 2nd ed. Wiley, New York. [2.1.5, 2.3.5, 2.3.7, 2.3.8, 3.5.3]

1 2

Cole, L. C. 1949. The measurement of interspecific association. *Ecology* 30: 411–424. [3.4.1]

1

Cole, L. C. 1957a. Biological clock in the Unicorn. *Science* 125: 874–876. [2.1.9]

1 2

Cole, L. C. 1957b. The measurement of partial interspecific association. *Ecology* 38: 226–233. [3.4.1]

2

Cole, L. C. 1962. A closed sequential test design for toleration experiments. *Ecology* 43: 749–753. [2.3.9, 4.2]

4

Coleman, M. J., and H. B. N. Hynes. 1970. The vertical distribution of the invertebrate fauna in the bed of a stream. *Limnol. Oceanog.* 15: 31–40. [2.3.6]

1 2

Colquhoun, D. 1971. *Lectures on biostatistics.* Clarendon Press. [2.1.2, 2.1.5, 2.1.8, 4.2]

1 2 3

Conway, G. R., N. R. Glass, and J. C. Wilcox. 1970. Fitting nonlinear models to biological data by Marquart's algorithm. *Ecology* 51: 503–507. [3.7, 3.10]

4

Cooley, J. M. 1977. Filtering rate performance of *Daphnia retrocurva* in pulp mill effluent. *J. Fish. Res. Bd. Can.* 34: 863–868. [3.7]

1 2 3 7

Cooley, W. W., and P. R. Lohnes. 1962. *Multivariate procedures for the behavioral sciences.* Wiley, New York. [2.3.9, 3.6, 3.10, 4.1, 4.4]

1 2 3 7

Cooley, W. W., and P. R. Lohnes. 1971. *Multivariate data analysis*. Wiley, New York. [2.3.9, 3.6, 3.7, 3.10, 3.11, 4.1, 4.3, 4.9]

2

Coons, J. 1957. The analysis of covariance as a missing plot technique. *Biometrics* **13**: 387–405. [3.5.3]

1 2

Cormack, R. M. 1968. The statistics of capture-recapture methods. *Oceanogr. Mar. Biol. Ann. Rev.* **6**: 455–506. [3.7]

2

Cormack, R. M. 1972. The logic of capture-recapture estimates. *Biometrics* **28**: 337–343. [3.7]

1

Council of Biology Editors, Committee on Form and Style. 1972. *CBE style manual*, 3rd ed. American Institute of Biological Sciences, Washington, D.C. [2.1.8]

1 4 5

Coutant, C. C. 1971. Thermal pollution—Biological effects. *J. Water Pollut. Control Fed.* **43**: 1292–1334. [3.7]

4

Cowell, B. C., and W. C. Carew. 1976. Seasonal and diel periodicity in the drift of aquatic insects in a subtropical Florida stream. *Freshwater Biol.* **6**: 587–594. [3.7]

2

Cox, D. R. 1957. Note on grouping. *J. Am. Stat. Assoc.* **52**: 543–547. [3.4.3]

1 2

Cox, D. R. 1970. *The analysis of binary data*. Methuen, London. [2.3.9, 3.4.1, 3.10]

1 7

Cox, D. R. 1972a. The analysis of multivariate binary data. *Appl. Stat.* **21**: 113–120. [3.4.1]

2

Cox, D. R. 1972b. Regression models and life-tables. *J. Roy. Stat. Soc. B* **34**: 187–220. [3.4.1]

2 7

Cox, D. R., and L. Brandwood. 1959. On a discriminatory problem connected with the works of Plato. *J. Roy. Stat. Soc. B* **21**: 195–200. [3.4.1]

1 2

Cox, D. R., and P. A. W. Lewis. 1966. *The statistical analysis of series of events*. Methuen, London. [4.2]

2

Cox, D. R., and A. Stuart. 1955. Some quick tests for trend in location and dispersion. *Biometrika* **42**: 80–95. [2.3.9]

7 9

Crawford, R. M. M., and D. Wishart. 1967. A rapid multivariate method for the detection and classification of groups of ecologically related species. *J. Ecol.* **55**: 505–524. [2.1.7, 4.4]

2 3 7

Crawford, R. M. M., and D. Wishart. 1968. A rapid classification and ordination method and its application to vegetation mapping. *J. Ecol.* **56**: 385–404. [4.4]

4 8

Crossman, J. S., and J. Cairns, Jr. 1974. A comparative study between two different artificial substrate samplers and regular sampling techniques. *Hydrobiologia* **44**: 517–522. [2.3.6, 3.5.2, 3.7]

4 7 8

Crossman, J. S., J. Cairns, Jr., and R. L. Kaesler. 1973. Aquatic invertebrate recovery in the Clinch River following pollutional stress. *Water Resour. Res. Cent. Bull*. 63, Virginia Polytechnic Institute and State University, Blacksburg, Va. [2.1.9]

4 7 8

Crossman, J. S., R. L. Kaesler, and J. Cairns, Jr. 1974. The use of cluster analysis in the assessment of spills of hazardous materials. *Am. Midl. Nat.* **92**: 94–114. [4.4]

1 2 7

Crovello, T. J. 1970. Analysis of character variation in ecology and systematics. *Ann. Rev. Ecol. Syst.* **1**: 55–98. [2.1.5, 2.1.7, 2.1.8, 3.6, 3.9, 3.11, 4.4]

5

Cunningham, P. A., and M. R. Tripp. 1973. Accumulation and depuration of mercury in the American oyster *Crassostrea virginica. Mar. Biol.* **20**: 14–19. [3.7]

5

Curtis, M. A. 1972. Depth distributions of benthic polychaetes in two fjords on Ellesmere Island, N.W.T. *J. Fish. Res. Bd. Can.* **29**: 1319–1327. [3.11]

4

Cvancara, A. M. 1963. Clines in three species of *Lampsilis* (Pelecypoda: Unionidae). *Malacalogia* **1**: 215–255. [2.1.9]

2 6 7

Dagnelie, P. 1965. L'étude des communautes vegetales par l'analyse statistique des liasons entre les especes et les variables ecologiques. *Biometrika* **21**: 345–361; 890–907. [4.4]

5

Dahlburg, M. D., and F. G. Smith. 1970. Mortality of estuarine animals due to cold on the Georgia coast. *Ecology* **51**: 931–933. [3.2]

2 7

Dale, M. B. 1964. The application of multivariate methods to heterogeneous data. Ph.D. thesis. University of Southampton, England. [2.3.9]

2 3 7

Dale, M. B., P. MacNaughton-Smith, W. T. Williams, and G. N. Lance. 1970. Numerical classification of sequences. *Aust. Comput. J.* **2**: 9–13. [4.2, 4.4]

2 3

Daniel, C., and F. S. Wood. 1971. *Fitting equations to data*. Wiley, New York. [2.1.7, 3.8, 3.9, 3.10]

2

Darroch, J. N. 1958. The multiple-recapture census. I. Estimation of a closed population. *Biometrika* **45**: 343–359. [3.7]

2

Darroch, J. N. 1961. The two-sample capture-recapture census when tagging and sampling are stratified. *Biometrika* **48**: 241–260. [3.7]

5

Davis, H. C., and H. Hidu. 1969. Effects of pesticides on embryonic development of clams and oysters and on survival and growth of the larvae. *Fish. Bull.* **67**: 393–404. [2.3.5, 3.7]

1 2 3 7

Davis, J. C. 1973. *Statistics and data analysis in geology*. Wiley, New York. [1.2, 3.10, 3.11, 4.2, 4.3]

2 7

Day, N. E. 1969. Estimating the components of a mixture of normal distributions. *Biometrika* **56**: 463–474. [4.4]

8

DeBenedictis, P. A. 1973. Correlations between certain diversity indices. *Am. Nat.* **107**: 295–302. [2.1.9]

8

Dejong, T. M. 1975. Comparison of three diversity indices based on their components of richness and evenness. *Oikos* **26**: 222–227. [3.5.2]

4 7 8

DeMarch, B. G. E. 1976. Spatial and temporal patterns in macrobenthic stream diversity. *J. Fish. Res. Bd. Can.* **33**: 1261–1270. [2.1.9, 2.3.3, 3.5.2, 3.5.4]

2

Dennison, J. M., and W. H. Hay. 1967. Estimating the needed sampling area for subaquatic ecologic studies. *J. Paleont.* **41**: 706–708. [3.8]

5

De Wilde, P. A. W. J. 1973. A continuous flow apparatus for long-term recording of oxygen uptake in burrowing invertebrates with some remarks on the uptake in *Macoma balthica*. *Neth. J. Sea Res.* **6**: 157–162. [3.7]

1 3

Dixon, W. J. 1950. Analysis of extreme values. *Ann. Math. Stat.* **21**: 488–506. [3.9]

3 7

Dixon, W. J. 1973. BMD: Biomedical computer programs. University of California, Berkeley, Calif. [3.5.3, 3.6, 3.7, 3.10, 3.11, 4.1, 4.2]

3 7

Dixon, W. J. 1975. BMDP: Biomedical computer programs. University of California, Berkeley, Calif. [3.10, 3.11]

6

Dunn, J. E., and P. S. Gipson. 1977. Analysis of radio telemetry data in studies of home range. *Biometrics* **33**: 85–101. [3.7, 3.11]

1 2 3 7

Duran, B. G., and P. L. Odell. 1974. Cluster analysis: A survey. Springer-Verlag, New York. [4.4]

2

Durbin, J. 1963. Trend elimination for the purpose of estimating seasonal and periodic components of time series. In *Proceedings of the symposium on time series analysis*, M. Rosenblatt, (Ed)., pp. 3–16. Wiley, New York. [4.2]

2 3

Ebert, T. A. 1973. Estimating growth and mortality rates from size data. *Oecologia* **11**: 281–298. [3.7, 3.10]

2 4

Edmonson, W. T. 1944. Ecological studies of sessile rotatoria. I. Factors affecting distribution. *Ecol. Monogr.* **14**: 31–36. [3.7]

7

Efron, B. 1975. Efficiency of logistic regression compared to normal discriminant analysis. *J. Am. Stat. Assoc.* **70**: 892–898. [3.4.1]

1 2 7

Eisenbeis, R. A., G. G. Gilbert, and R. B. Avery. 1973. Investigating the relative importance of individual variables and variable subsets in discriminant analysis. *Commun. Stat.* **2**: 205–219. [3.6]

1

Eisenhart, C. 1947. The assumptions underlying the analysis of variance. *Biometrics* **3**: 1–21. [2.1.7]

1

Eisenhart, C. 1968. Expression of the uncertainties of final results. *Science* **160**: 1201–1204. [2.1.5, 2.1.8]

1 7

Elashoff, R. M. and A. A. Afifi. 1966. Missing observations in multivariate statistics. I Review of the literature. *J. Am. Stat. Assoc.* **61**: 595–601. [3.5.3]

1 2 4

Elliott, J. M. 1977. Some methods for the statistical analysis of samples of benthic invertebrates. *Sci. Publ.* No. 25, Freshwater Biological Association, Ferry House, U.K. [2.1.1, 2.1.5, 2.1.7, 2.3.5, 2.3.8, 2.3.9, 3.8]

1 2 4

Elliott, J. M., and H. Décamps. 1973. Guide pour l'analyse statistique des échatillons d'invertebrés benthiques. *Ann. Limnol.* **9**: 79–120. [2.1.1, 2.3.8]

3 5

Ellis, D. V. 1968. A series of computer programs for summarizing data from quantitative benthic investigations. *J. Fish. Res. Bd. Can.* **25**: 1737–1738. [3.4.2, 3.10]

8

Engstrom-Heg, V. L. 1970. Predation, competition and environmental variables: Some mathematical models. *J. Theor. Biol.* **27**: 175–195. [3.5.2]

1 3 7

Enslein, K., A. Ralston, and H. S. Wilf, (Eds.), 1977. *Statistical methods for digital computers.* Vol. 3 of *Mathematical methods for digital computers.* Wiley, New York. [3.10]

1 4 5 7

Erman, D. C., and W. T. Helm. 1971. Comparison of some species importance values and ordination techniques used to analyze benthic invertebrate communities. *Oikos* **22**: 240–247. [3.5.1, 4.4]

5

Evans, R. J., J. D. Bails, and F. M. D'Itri. 1972. Mercury levels in muscle tissues of preserved marine fish. *Environ Sci.* Technol. **6**: 901–905. [3.7, 4.5]

2 8 9

Fager, E. W. 1972. Diversity: A sampling study. *Am. Nat.* **106**: 293–310. [2.1.7, 3.5.2]

3 7 9

Fasham, M. J. R. 1977. A comparison of nonmetric multidimensional scaling, principal components and reciprocal averaging for the ordination of simulated coenoclines, and coenoplanes. *Ecology* **58**: 551–561. [3.4.2, 4.4]

5

Fenchel, T. 1975. Factors determining the distribution patterns of mud snails (Hydrobiidae). *Oecologia* **20**: 1–17. [2.1.9]

1 2 7

Fienberg, S. E. 1970. The analysis of multidimensional contingency tables. *Ecology* **51**: 419–433. [2.3.9, 3.4.1, 3.4.3]

5

Fincham, A. A. 1971. Ecology and population studies of some intertidal and sublittoral sand-dwelling amphipods. *J. Mar. Biol. Assoc. U.K.* **51**: 471–488. [3.11]

1 2

Finney, D. J. 1964. Probit analysis. *A statistical treatment of the sigmoid response curve*. Cambridge University Press. [4.2]

3 6 7

Fisher, D. R. 1968. A study of faunal resemblance using numerical taxonomy and factor analysis. *Syst. Zool.* **17**: 48–63. [3.4.1, 3.11]

8

Fisher, R. A., A. G. Corbett, and C. B. Williams. 1943. The relation between number of species and the number of individuals in a random sample of an animal population. *J. Anim. Ecol.* **12**: 42–58. [3.5.2]

1

Fisher, R. A., and F. Yates. 1967. *Statistical tables for biological, agricultural and medical research*. Oliver and Boyd, Edinburgh. [2.3.9]

4

Flannagan, J. F. 1973a. Sorting benthos using flotation media. *Fish. Res. Bd. Can. Techn. Rep.* No. 354, Freshwater Institute, Winnipeg. [2.3.6]

4

Flannagan, J. F. 1973b. Field and laboratory studies of the effect of exposure to fenitrothion on freshwater aquatic invertebrates. *Manitoba Entomol.* **7**: 15–25. [3.7]

5

Fleminger, A., and R. I. Clutter. 1965. Avoidance of towed nets by zooplankton. *Limnol. Oceanogr.* **10**: 96–104. [2.3.6]

5

Foehrenbach, J. 1971. Cholorinated hydrocarbon residues in shellfish (Pelecypoda) from estuaries of Long Island, N.Y. *Pestic. Monit. J.* **5**: 242–247. [3.7]

5 7 9

Fox, W. T. 1969. Analysis and simulation of paleoecologic communities through time. *Proc. N. A. Paleontol. Conv. B*; 117–135. [4.2]

1 7

Frane, J. W. 1976. Some simple procedures for handling missing data in multivariate analysis. *Psychometrika* **41**: 409–415. [3.5.3]

5

Frank, P. W. 1965. Shell growth in a natural population of the turban shell, *Tegula funebralis*. *Growth* **29**: 395–403. [3.7]

2 7

Fraser, A. R., and M. Kovats. 1966. Stereoscopic models of multivariate statistical data. *Biometrics* **22**: 358–367. [3.11]

2 7

Freeman, H., and A. M. Kuzmack. 1972. Tables of multivariate *t* in six and more dimensions. *Biometrika* **59**: 217–219. [3.4.2]

2 7

Freeman, H., A. M. Kuzmack, and R. Maurice. 1967. Multivariate *t* and the ranking problem. *Biometrika* **54**: 305. [3.4.2]

2 7

Friedman, H. P., and J. Rubin. 1967. On some invariant criteria for grouping data. *J. Am. Stat. Assoc.* **62**: 1159–1178. [3.4.2]

5 7

Gage, J. 1972. A preliminary survey of the benthic macrofauna and sediments in Lochs Etive and Creran, sea-lochs along the west coast of Scotland. *J. Mar. Biol. Assoc. U.K.* **52**: 237–276. [2.3.6, 3.11]

3

Gallucci, V. F., and J. Hylleberg. 1976. Quantification of some aspects of growth in bottom-feeding bivalve *Macoma nasuta. Veliger* **19**: 59–67. [3.7]

3 7 9

Gauch, H. G., Jr. 1976. *Catalog of the Cornell Ecology Programs Series.* Ecology and Systematics, Cornell University, Ithaca, N.Y. [2.1.7, 3.4.2, 3.10]

2 3

Gauch, H. G., Jr., and G. B. Chase. 1974. Fitting the Gaussian curve to ecological data. *Ecology* **55**: 1377–1381. [4.4]

1 6 7

Gauch, H. G., Jr., and R. H. Whittaker. 1972. Comparison of ordination techniques. *Ecology* **53**: 868–875. [4.4]

2 3 7

Gauch, H. G., Jr., G. B. Chase, and R. H. Whittaker. 1974. Ordination of vegetation samples by Gaussian species distributions. *Ecology* **55**: 1382–1390. [4.4]

1 7 9

Gilbert, E. S. 1968. On discrimination using qualitative variables. *J. Am. Stat. Assoc.* **63**: 1399–1418. [3.4.3]

3 6 7

Gittins, R. 1968. Trend-surface analysis of ecological data. *J. Ecol.* **56**: 845–869. [3.11]

2 3

Glass, N. R. 1967. A technique for fitting nonlinear models to biological data. *Ecology* **48**: 1010–1013. [3.7, 3.10]

1 9

Glass, G. V., P. D. Peckham, and J. R. Sanders. 1972. Consequences of failure to meet assumptions underlying the fixed effects analyses of variance and covariance. *Rev. Educ. Res.* **42**: 237–288. [2.1.7, 2.3.3, 2.3.9, 3.4.3, 3.5.3]

1 7

Gnanadesikan, R., and J. R. Kettenring. 1972. Robust estimates, residuals and outlier detection with multiresponse data. *Biometrics* **28**: 81–124. [3.9[

2 6 7

Goodall, D. W. 1961. Objective methods for the classification of vegetation. IV. Pattern and minimal area. *Aust. J. Bot.* **9**: 162–196. [3.8]

1 3 6 7 8

Goodall, D. W. 1962. Bibliography of statistical plant sociology. *Excerpta Bot. Ser. B*, **4**: 253–322. [3.5.2]

7

Goodall, D. W. 1966a. Hypothesis testing in classification. *Nature, Lond.* **211**: 329–330. [2.1.2]

2 7

Goodall, D. W. 1966b. A new similarity index based on probability. *Biometrics* **22**: 882–907. [3.4.3, 3.5.4, 3.9, 4.4]

7

Goodall, D. W. 1966c. Numerical taxonomy of bacteria—Some published data re-examined. *J. Gen. Microbiol.* **42**: 25–37. [3.4.3, 3.5.4, 3.9]

2 3 9

Goodall, D. W. 1967. The distribution of the matching coefficient. *Biometricx* **23**: 647–656. [3.4.1]

1 7

Goodall, D. W. 1973a. Sample similarity and species correlation. In *Ordination and classification of communities*. Part V, *Handbook of vegetation science*, R. H. Whittaker (ed.), pp. 105–156. W. Junk, The Hague. [3.4.1]

1 7

Goodall, D. W. 1973b. Numerical methods of classification. In *Ordination and classification of communities*, Part V, *Handbook of vegetation science*, R. H. Whittaker, (ed.), pp. 575–618. W. Junk, The Hague. [3.4.2, 3.4.3, 3.8, 3.9, 4.4]

1 8

Goodman, D. 1975. The theory of diversity-stability relationships in ecology. *Q. Rev. Biol.* **50**: 237–266. [3.5.2]

2 7

Gower, J. C. 1966a. Some distance properties of latent root and vector methods used in multivariate analysis. *Biometrika* **53**: 325–338. [3.4.3, 3.9]

1 7

Gower, J. C. 1967a. Multivariate analysis and multidimensional geometry. Statistician **17**: 13–28. [2.3.9, 3.4.2, 3.4.3, 3.9, 3.11]

1 7

Gower, J. C. 1967b. A comparison of some methods of cluster analysis. *Biometrics* **23**: 623–637. [3.4.3, 4.4]

5

Grassle, J. P., and J. F. Grassle. 1976. Sibling species in the marine pollution indicator *Capitella* (Polychaeta). *Science* **192**: 567–569. [3.6]

5

Grassle, J. F., H. L. Sanders, R. R. Hessler, G. T. Rowe, and T. McLellan. 1975. Pattern and zonation: a study of the bathyl megafauna using the research submersible Alvin. *Deep-Sea Res.* **22**: 457–481. [2.3.7]

2 4

Green, R. H. 1965. Estimation of tolerance over an indefinite time period. *Ecology* **46**: 887. [4.2]

5

Green, R. H. 1968. Mortality and stability in a low-diversity sub-tropical intertidal community. *Ecology* **49**: 848–854. [2.3.7, 3.5.3]

2

Green, R. H. 1970a. Graphical estimation of rates of mortality and growth. *J. Fish. Res. Bd. Can.* **27**: 204–208. [3.7, 3.10]

2

Green, R. H. 1970b. On fixed precision level sequential sampling. *Res. Popul. Ecol.* **12**: 249–251. [3.8]

2 4 7

Green, R. H. 1971a. A multivariate statistical approach to the Hutchinsonian niche: Bivalve molluscs of central Canada. *Ecology* **52**: 543–556. [2.1.9, 2.3.9]

4

Green, R. H. 1971b. Lipid and caloric contents of the relict amphipod *Pontoporeia affinis* in Cayuga Lake, N.Y. *J. Fish. Res. Bd. Can.* **28**: 776–777. [3.7]

4 7

Green, R. H. 1972. Distribution and morphological variation of *Lampsilis radiata* (Pelecypoda, Unionidae) in some central Canadian lakes: A multivariate statistical approach. *J. Fish. Res. Bd. Can.* **29**: 1565–1570. [3.7]

5

Green, R. H. 1973. Growth and mortality in an arctic intertidal population of *Macoma balthica* (Pelecypoda, Tellinidae). *J. Fish. Res. Bd. Can.* **30**: 1345–1348. [3.7]

2 4 7

Green, R. H. 1974. Multivariate niche analysis with temporally varying environmental factors. *Ecology* **55**: 73–83. [3.6, 4.4]

2 3 4 7 9

Green, R. H. 1977. Some methods for hypothesis testing and analysis with biological monitoring data. In *Biological Monitoring of water and effluent quality*, ASTM STP 607, J. Cairns, Jr., K. L. Dickson, and G. F. Westlake, (Eds.), pp. 200–211. American Society for Testing and Materials. [2.1.7, 3.5.2, 4.1, 4.2, 4.4]

2 3 4 7 9

Green, R. H. In press. Optimal impact study design and analysis. In *Biological data in water quality assessment*, K. L. Dickson, J. Cairns, Jr., and R. J. Livington (Eds.), American Society for Testing and Materials STP, Philadelphia. [4.1]

5 7

Green, R. H., and K. D. Hobson. 1970. Spatial and temporal structure in a temperate intertidal community, with special emphasis on *Gemma gemma* (Pelecypoda Mollusca). *Ecology* **51**: 999–1011. [2.3.6, 2.3.7, 2.3.9, 3.5.3, 3.7, 3.11, 4.4]

1 2 3 4 7 8

Green, R. H., and G. L. Vascotto. In press. Analysis of environmental factors controlling spatial patterns in species composition. *Water Res.* [4.4]

1 2 6 7 8

Greig-Smith, P. 1964. *Quantitative plant ecology*. Butterworths, London. [2.3.7, 2.3.8, 3.4.1]

2 7

Grizzle, J. E., and O. D. Williams. 1972. Log linear models and tests of independence for contingency tables. *Biometrics* **28**: 137–156. [3.5.3]

1

Gruenberger, F. J. 1964. A measure for crackpots. *Science* **145**: 1413–1415. [2.1.5, 2.1.8]

8

Hairston, N. G. 1964. Studies on the organization of animal communities. Jubilee symposium supplement, *J. Ecol.* **52**: 227–239. [3.5.2]

3 7

Hall, D. J., G. H. Ball, D. E. Wolf, and J. W. Eusebio. 1968. Promenade—An interactive graphics pattern-recognition system. *Stanford Res. Inst. Rep.* Menlo Park, Calif. [3.10]

2 4 9

Hamilton, A. L. 1969. On estimating annual production. *Limnol. Oceanogr.* **14**: 771–782. [3.7]

4

Hamilton, A. L., W. Burton, and J. F. Flannagan. 1970. A multiple corer for sampling profundal benthos. *J. Fish. Res. Bd. Can.* **27**: 1867–1869. [2.3.7]

2

Hamon, B. V., and E. J. Hannan. 1963. Estimating relations between time series. *J. Geophys. Res.* **68**: 6033–6041. [2.1.9, 4.2]

5

Hamwi, A., and H. H. Haskin. 1969. Oxygen consumption and pumping rates in the hard clam *Mercenaria mercenaria*: A direct method. *Science* **163**: 823–824. [3.7]

1 2

Hannan, E. J. 1960. *Time series analysis*. Methuen, London. [4.2]

1 2 7

Hannan, E. J. 1970. *Multiple time series*. Wiley, New York. [4.2]

4 7 8

Harman, W. N. 1972. Benthic substrates: Their effect on freshwater mollusca. *Ecology* **53**: 271–277. [3.5.2]

1 2 3 7

Harris, R. J. 1975. *A primer of multivariate statistics*. Academic Press, New York. [2.3.9, 3.4.3, 3.7, 3.9, 4.1, 4.2]

1 4

Hart, C. W., Jr., and S. L. H. Fuller, (Eds.), 1974. *Pollution ecology of freshwater invertebrates*. Academic Press, New York [3.6]

6

Haugen, R. K. 1967. Tree ring indices: A circumpolar comparison. *Science* **158**: 773–775. [3.7, 4.5]

1

Healy, M. J. R., and L. R. Taylor. 1962. Tables for power-law transformations. *Biometrika* **49**: 557–559. [2.3.9]

2 8

Heck, K. L., G. van Belle, and D. Simberloff. 1975. Explicit calculation of rarefaction diversity measurement and determination of sufficient sample size. *Ecology* **56**: 1459–1461. [3.5.2]

1 8

Hedgpeth, J. W. 1977. Models and muddles: Some philosophical observations. *Helgö-lander Wiss. Meeresunters.* **30**: 92–104 [2.1.4]

4 5 7 8

Hendricks, A., D. Henley, J. T. Wyatt, K. L. Dickson, and J. K. Silvey. 1974. Utilization of diversity indices in evaluating effect of a paper mill effluent on bottom fauna. *Hydrobiologia* **44**: 463–474. [3.2, 3.5.2]

4 7 9

Herricks, E. E., and J. Cairns, Jr. 1974. The recovery of streams stressed by acid coal mine drainage. Fifth Symposium on Coal Mine Drainage Research, Louisville, Ky. [2.1.7]

6

Hickey, J. J., and D. W. Anderson. 1968. Chlorinated hydrocarbons and eggshell changes in raptorial and fish-eating birds. *Science* **162**: 271–273. [3.7, 4.5]

4 5

Hicks, D. B., and J. W. DeWitt. 1970. A system for maintaining constant dissolved oxygen concentrations in flowing-water experiments. *Prog. Fish. Cult.* **32**: 55–57. [3.7]

3 9

Hilborn, R. 1973. A control system for FORTRAN simulation programming. *Simulation* 172–175. [2.1.7]

2 7

Hills, M. 1969. On looking at large correlation matrices. *Biometrika* **56**: 249–253. [3.9]

6

Hinckley, A. D. 1969. Radiation-induced fluctuations in forest insect populations. *Rad. Res.* **39**: 502. [2.1.2]

2 3 4 7 8

Hocutt, C. H., R. L. Kaesler, M. T. Masnik, and J. Cairns, Jr. 1974. Biological assessment of water quality in a large river system: An evaluation of a method for fishes. *Arch. Hydrobiol.* **74** : 448–462. [3.10]

Hoffer, E. 1951. *The true believer*. Harper and Row, New York. [3.5.2]

5 7 8

Holland, A. F., and T. T. Polgar. 1976. Seasonal changes in structure of an intertidal community. *Mar. Biol.* **37**: 341–348. [3.5.2]

1 5

Holme, N. A., and A. D. McIntyre, (Eds.), 1971. Methods for the study of marine benthos. *Blackwell Sci. Publ., Oxford. IBP Handb.* **16**: 334. [3.7]

1 2 7

Hope, K. 1968. *Methods of multivariate analysis*. University of London Press, London. [3.6]

2 7

Horton, J. F., J. S. Russell and A. W. Moore. 1968. Multivariate convariance and canonical analysis: A method for selecting the most effective discriminators in a multivariate situation. *Biometrics* **24**: 845–858. [3.6]

5

Howell, B. R., and R. G. J. Shelton. 1970. The effect of China clay on the bottom fauna of St. Austell and Mevagissey Bays. *J. Mar. Biol. Assoc. U.K.* **50**: 593–607. [3.11]

5 7

Hughes, R. N. and M. L. H. Thomas. 1971. Classification and ordination of benthic samples from Bedeque Bay, an estuary in Prince Edward Island, Canada. *Mar. Biol.* **10**: 227–235. [2.3.9, 4.4]

5 7

Hughes, R. N., and M. L. H. Thomas. 1971. The classification and ordination of shallow-water benthic samples from Prince Edward Island, Canada. *J. Exp. Mar. Biol. Ecol.* **7**: 1–39. [2.1.8]

5

Hummon, W. D. 1974. Effects of DDT on longevity and reproductive rate in *Lepidodermella squammata* (Gastrotricha, Chaetonotida). *Am. Midl. Nat.* **92**: 327–339. [4.2]

5

Hummon, W. D., and M. R. Hummon. 1975. Use of life table data in tolerance experiments. *Cah. Biol. Mar.* **16**: 743–740. [4.2]

4

Hunt, P. C., and J. W. Jones. 1972. The effect of water level fluctuations on a littoral fauna. *J. Fish. Biol.* **4**: 385–394. [3.11]

Huntington, E. 1945. *Mainsprings of civilization*. New American Library, New York. [2.1.9]

6 8

Hurd, L. E., M. V. Mellinger, L. L. Wolf, and S. J. McNaughton. 1971. Stability and diversity at three trophic levels in terrestrial successional ecosystems. *Science* **173**: 1134–1136. [3.5.2]

2 9

Hurlbert, S. H. 1969. A coefficient of interspecific association. *Ecology* **50**: 1–9. [2.1.7, 3.4.1]

1 2 8

Hurlbert, S. H. 1971. The nonconcept of species diversity: A critique and alternative parameters. *Ecology* **52**: 577–586. [3.5.2, 3.7, 4.2]

1 4

Hurlbert, S. H. 1975. Secondary effects of pesticides on aquatic ecosystems. *Residue Rev.* **58**: 81–148. [2.1.1, 2.1.8, 2.3.4]

Hutchinson, G. E. 1968. When are species necessary? In *Population biology and evolution*, R. C. Lewontin, (ed.), pp. 177–186. Syracuse University Press, Syracuse, N.Y. [2.1.9]

4

Hynes, H. B. N., D. D. Williams, and N. E. Williams. 1976. Distribution of the benthos within the substratum of a Welsh mountain stream. *Oikos* **27**: 307–310. [2.3.6]

2 6 7 9

Ihm, P., and H. van Groenewoud. 1975. A multivariate ordering of vegetation data based on Gaussian type gradient response curves. *J. Ecol.* **63** : 767–777. **[4.4]**

3

International Business Machines Corp. 1970. *System/360 scientific subroutine package, Version III programmers guide*, White Plains, N.Y. [3.10]

6 7

Ivimey-Cook, R. B., and M. C. F. Proctor. 1967. Factor analysis of data from an East Devon heath: A comparison of principal component and rotated solutions. *J. Ecol.* **55**: 405–413 [3.4.3]

6 7

James, F. 1971. Ordinations of habitat relationships among breeding birds. *Wilson Bull.* **83**: 215–236. [3.11, 4.4]

1 2 7

Jardine, N., and R. Sibson. 1968. The construction of hierarchic and non-hierarchic classifications. *Comput. J.* **11**: 177–184. [4.4]

1 2 3 7

Jardine, N., and R. Sibson. 1971. *Mathematical taxonomy*. Wiley, London. [3.10, 4.4]

2 7

Jensen, D. R. 1972. Some simultaneous multivariate procedures using Hotelling's T^2 statistics. *Biometrics* **28**: 39–53. [4.2]

4

Johnson, M. G. 1977. Caloric changes along pulp and paper mill effluent plumes. *J. Fish. Res. Bd. Can.* **34**: 784–790. [2.3.7]

2 7

Jolicoeur, P. 1963a. Bilateral symmetry and asymmetry in limb bones of *Martes americana* and man. *Rev. Can. Biol.* **22**: 409–432. [2.3.9]

2 7

Jolicoeur, P. 1963b. The degree of generality of robustness in *Martes americana*. *Growth* **27**: 1–27. [2.3.9]

2 7

Jolicoeur, P. and J. E. Mosimann. 1960. Size and shape variation in the painted turtle. A principal component analysis. *Growth* **24**: 339–354. [2.3.9, 3.7, 4.3]

2

Jolly, G. M. 1965. Explicit estimates from capture-recapture data with both death and immigration—Stochastic model. *Biometrika* **52**: 225–247. [3.7]

3 7

Jones, K. J. 1964. *The multivariate statistical analyser* (A system of Fortran II programs to be run under 7090-4 Fortran monitor system). Harvard (Education). [2.1.6, 3.10]

2 7

Jones, R. H., D. H. Cromwell, and L. E. Kapuniai. 1970. Change detection model for serially correlated multivariate data. *Biometrics* **26**: 269–280. [4.2]

4 7

Kaesler, R. L., J. Cairns, Jr., and J. S. Crossman. 1974. Redundancy in data from stream surveys. *Water Res.* **8**: 637–642. [3.6]

4

Kaiser, K. L. E. 1977. Organic contaminant residues in fishes from Nipigon Bay, Lake Superior. *J. Fish. Res. Bd. Can.* **34**: 850–855. [3.7]

4

Kamp-Nielsen, L. 1971. The effect of deleterious concentrations of mercury on the photosynthesis and growth of *Chlorella pyrenoidosa*. *Physiol. Plant.* **24**: 556–561. [3.7]

1 4

Katz, A. 1972. Mercury pollution: The making of an environmental crisis. *Crit. Rev. Environ. Control* **2**: 517–534. [3.6, 3.7, 4.5]

5

Keckes, S., B. Ozretić, and M. Krajnović. 1969. Metabolism of Zn^{65} in mussels (*Mytilus galloprovincialis* Lam.). Uptake of Zn^{65}. *Rapp. Comm. Int. Mer Medit.* **19**: 949–952. [3.7]

4

Kelso, J. R. M. 1977. Density, distribution, and movement of Nipigon Bay fishes in relation to a pulp and paper mill effluent. *J. Fish. Res. Bd. Can.* **34**: 879–885. [3.7]

4

Kelso, J. R. M., C. K. Minna, and R. J. P. Brouzeo. 1977. Pulp and paper mill effluent in a freshwater environment. *J. Fish. Res. Bd. Can.* **34**: 771–775. [3.6]

2

Kendall, D. G. 1975. The recovery of structure from fragmentary information. *Phil. Trans. Roy. Soc. A* **279**: 547–582. [4.4]

2 7

Kendall, M. G. 1966. Discrimination and classification. In *Multivariate analysis*, P. R. Krishnaiah, (Ed.). Academic Press, New York. [3.6, 4.1, 4.4]

1

Kendall, M. G., and A. Stuart. 1969. *The advanced theory of statistics*, Vol. 1; *Distribution theory*. Hafner, New York. [3.5.3]

5

Kennedy, V. S. and J. A. Mihursky. 1971. Upper temperature tolerances of some estuarine bivalves. *Chesapeake Sci.* **12**: 193–204. [4.2]

5

Kennedy, V. S., and J. A. Mihursky. 1972. Effects of temperature on the respiratory metabolism of three Chesapeake Bay bivalves. *Chesapeake Sci.* **13**: 1–22. [3.7]

1 4 5 7

Keup, L. E., W. M. Ingram, and K. M. Mackenthun. 1966. *The role of bottom dwelling macrofauna in water pollution investigations*. U.S. Department of Health, Education and

Welfare, Public Health Service, Division of Water Supply and Pollution Control. 23 pp. [3.4.1, 3.6]

37

Klecka, W. R. 1975. Discriminant analysis. In *Statistical package for the social sciences*, 2nd ed. McGraw-Hill, New York. [4.1]

2

Knight, W. 1974. A run-like statistic for ecological transects. *Biometrics* **30**: 553–556. [3.4.1]

4

Kobylinski, G. J., and R. J. Livingston. 1975. Movement of mirex from sediment and uptake by the hogchoker, *Trinectes maculatus*, *Bull. Environ. Contamination Toxicol.* **14**: 692–698. [2.3.4, 3.7]

7

Korin, B. P. 1972. Some comments on the homoscedasticity criterion, M, and the multivariate analysis of variance tests T^2, W and R. *Biometrika* **59**: 215–217. [4.1]

Körner, S. 1966. *Experiences and theory*. Routledge and Kegan Paul, London. [2.1.2]

27

Kruskal, J. B. 1964a. Multidimensional scaling by optimizing goodness of fit to a nonmetric hypothesis. *Psychometrika* **29**: 1–27. [3.4.2]

237

Kruskal, J. B. 1964b. Non-metric multidimensional scaling: A numerical method. *Psychometrika* **29**: 115–129. (3.4.2]

37

Kruskal, J. B. 1977. Multidimensional scaling and other methods for discovering structure. In *Statistical methods for digital computers*, Vol. 3 of *Mathematical methods for digital computers*, K. Enslein, A. Ralston, and H. S. Wilf (eds.), pp. 296–339, Wiley, New York. [3.4.2]

27

Krzanowski, W. J. 1971. A comparison of some distance measures applicable to multinomial data, using a rotational fit technique. *Biometrics* **27**: 1062. [3.4.1]

178

Kullback, S. 1968. *Information theory and statistics*. Dover, New York. [3.5.2]

2

Kuno, E. 1969. A new method of sequential sampling to obtain the population estimates with a fixed level of precision. *Res. Popul. Ecol.* **11**: 127–136. [3.8]

2

Kuno, E. 1972. Some notes on population estimation by sequential sampling. *Res. Popul. Ecol.* **14**: 58–73. [2.3.9]

57

Kuris, A. M., and M. S. Brody. 1976. Use of principal components analysis to describe snail shell resource for hermit crabs. *J. Exp. Mar. Biol. Ecol.* **22**: 69–77. [3.5.3]

7

Lachenbruch, P. A., and L. L. Kupper. 1973. Discriminant analysis when one population is a mixture of normals. *Biom. Z.* **15**: 191–197. [3.5.3]

45

Lackey, R. T., and B. E. May. 1971. Use of sugar flotation and dye to sort benthic samples. *Trans. Am. Fish. Soc.* **100**: 794–797. [2.3.6]

7 9

LaFrance, C. R. 1972. Sampling and ordination characteristics of computer-simulated individualistic communities. *Ecology* **53**: 387–397. [2.1.7, 2.3.8, 4.4]

3 6 7

Lambert, J. M. 1972. Theoretical models for large-scale vegetation survey. In *Mathematical models in ecology*, J. N. R. Jeffers, (Ed.), pp. 87–109. Oxford, Blackwell. [2.3.1, 2.3.7]

1 6 7

Lambert, J. M., and M. B. Dale. 1964. The use of statistics in phytosociology. *Adv. Ecol. Res*. **2**: 59–99. [3.4.1, 3.4.3, 4.4]

6 7

Lambert, J. M., and W. T. Williams. 1966. Multivariate methods in plant ecology. VI. Comparison of information analysis with association analysis. *J. Ecol*. **54**: 635–664. [2.1.7]

1 3 7

Lance, G. N., and W. T. Williams. 1967a. Mixed data classification programs. I. Agglomerative systems. *Aust. Comput. J*. **1**: 1–6. [3.4.3]

1 3 7

Lance, G. N., and W. T. Williams. 1967b. A general theory of classificatory sorting strategies. II. Clustering systems. *Comput. J*. **10**: 271–277. [4.4]

1 3 7

Lance, G. N., and W. T. Williams. 1968. Mixed-data classificatory programs. II. Divisive systems. *Aust. Comput. J*. **1**: 82–85. [4.4]

1 2 7

Lazarsfeld, P. F., and N. W. Henry. 1968. *Latent structure analysis*. Mifflin, Boston. [3.4.1]

3

Lee, E. T. 1974. A computer program for linear logistic regression analysis. *Comput. Prog. Biomed*. **4**: 80–92. [3.4.1, 3.10]

4

Lee, F. G., and W. Wilson. 1969. Use of chemical composition of freshwater clamshells as indicators of paleohydrologic conditions. *Ecology* **50**: 990–997. [3.7]

1 2 3 7

Lee, P. J. 1971. Multivariate analysis for the fisheries biology. *Fish. Res. Bd. Can. Tech. Rep*. No. 244, Freshwater Institute, Winnipeg, Canada. [2.3.8, 3.10]

5

Lee, R. F., R. Sauerheber, and A. A. Benson. 1972. Petroleum hydrocarbons: Uptake and discharge by the marine mussel *Mytilus edulis. Science* **177**: 344–346. [3.7]

2 7

Lefkovitch, L. P. 1976. Hierarchical clustering from principal coordinates: An efficient method for small to very large numbers of objects. *Math. Biosci*. **31**: 157–174. [3.4.1, 3.5.3, 3.9, 4.4]

5 7

Legendre, L. 1973. Phytoplankton organization in Baie des Chaleurs (Gulf of St. Lawrence). *J. Ecol*. **61**: 135–149 [4.4]

4

Lehmkuhl, D. M. 1972. Change in thermal regime as a cause of reduction of benthic fauna downstream of a reservoir. *J. Fish. Res. Bd. Can*. **29**: 1329–1332. [3.7]

6

Leonard, D. E. 1970. Intrinsic factors causing qualitative changes in populations of *Por-thetria dispar* (Lepidoptera; Lymantrudae). *Can. Entomol.* **102**: 239–249. [2.1.9]

4

Leslie, J. K. 1977. Characterization of suspended particles in some pulp and paper mill effluent plumes. *J. Fish. Res. Bd. Can.* **34**: 791–797. [2.3.7]

4 7

Levandowsky, M. 1972. An ordination of phytoplankton populations in ponds of varying salinity and temperature. *Ecology* **53**: 398–407. [3.4.2]

4

Lévêque, Ch. 1971. Équation de von Bertalanffy et croissance des mollusques benthiques du Lac Tchad. *Cah. O.R.S.T.O.M., Hydrobiol.* **3/4**: 263–283. [3.7]

1

Levins, R. 1966. The strategy of model building in population biology. *Am. Sci.* **54**: 421–431. [2.1.4, 2.1.7, 2.3.9]

5

Levinton, J. S., and R. K. Bambach. 1970. Some ecological aspects of bivalve mortality patterns. *Am. J. Sci.* **268**: 97–112. [3.7]

1

Lewis, T., and L. R. Taylor. 1967. *Introduction to experimental ecology*. Academic Press, London. [2.1.8, 3.11]

8

Lewontin, R. 1969. The meaning of stability. In *Diversity and stability of ecological systems*. Brookhaven Symposium on Biology, Vol. 22, pp. 13–24. [3.5.2]

2 4 5 6

Lieth, H., and R. Whittaker, (Eds.). 1975. *Primary productivity of the biosphere*. Springer-Verlag, New York. [3.11]

4 5

Livingston, R. J. 1968. A volumetric respirometer for long-term studies of small aquatic animals. *J. Mar. Biol. Assoc. U.K.* **48**: 485–497. [3.7]

4 5

Livingston, R. J. 1970. A device for the continuous respirometry of small aquatic animals. *Copeia* **4**: 756–758. [3.7]

5 7 8

Livingston, R. J. 1975. Impact of kraft pulp-mill effluent on estuarine and coastal fishes in Apalachee Bay, Florida, U.S.A. *Mar. Biol.* **32**: 19–48. [3.5.2]

5 7 8

Livingston, R. J. 1976. Diurnal and seasonal fluctuations of organisms in a North Florida estuary. *Estuarine Coastal Mar. Sci.* **4**: 373–400. [4.2]

1 5 7 8

Livingston, R. J. 1977. Time as a factor in biomonitoring estuarine systems with reference to benthic macrophytes and epibenthic fishes and invertebrates. *Biological monitoring of water and effluent quality*, ASTM STP607. J. Cairns, Jr., K. L. Dickson, and G. F. Westlake, (Eds.), pp. 212–234. American Society for Testing and Materials. [3.8, 4.2]

5 8

Livingston, R. J., R. S. Lloyd, and M. S. Zimmerman. 1976. Determination of sampling strategy for benthic macrophytes in polluted and unpolluted coastal areas. *Bull. Mar. Sci.* **26**: 569–575. [3.8]

5

Livingston, R. J., C. R. Cripe, C. C. Koenig, F. G. Lewis III, and B. D. DeGrove. 1974. A system for the determination of chronic effects of pollutants on the physiology and behavior of marine organisms. *State Univ. Syst. Fla. Sea Grant Programs. Rep.* 4: 15. [3.7]

28

Lloyd, M., and R. J. Ghelardi. 1964. A table for calculating the "equitability" component of species diversity. *J. Anim. Ecol.* 33: 217–225. [3.5.2]

6

Loucks, O. L. 1962. Ordinating forest communities by means of environmental scalars and physotociological indices. *Ecol. Monogr.* 32: 137–166. [2.3.9]

8

MacArthur, R. 1955. Fluctuations of animal populations, and a measure of community stability. *Ecology* 36: 533–536. [3.5.2]

3 5 7 8

Macdonald, K. B. 1969. Quantitative studies of salt marsh faunas from the North American Pacific coast. *Ecol. Monogr.* 39: 33–60. [3.5.2]

4 8

Mackay, R., and J. Kalff. 1969. Seasonal variation in standing crop and species diversity of insect communities in a small Quebec stream. *Ecology* 50: 101–109 [3.5.2]

2 3 7

MacNaughton-Smith, P. 1963. The classification of individuals by the possession of attributes associated with a criterion. *Biometrics* 19: 364–366. [3.4.1, 4.1, 4.4]

9

Maelzer, D. A. 1970. The regression of log N_{n+1} on log N_n as a test of density-dependence: An exercise with computer constructed density-dependent populations. *Ecology* 51: 810–822. [2.1.7]

7

Mager, P. P. 1974. Analysis of multivariate data—Variance analysis in time series. *Act. Nerv. Super.* 16: 141–142. [4.1]

2 7 9

Maile, M. H. 1972. Randomization test for multivariate data. *Biometrics* 28: 269. [2.1.7, 4.4]

2 7

Maloney, C. J. 1974. Combining multiple attribute outcomes into an overall index. In *Statistical and mathematical aspects of pollution problems*, J. W. Pratt, (Ed.). Marcel Dekker, New York [3.5.1]

2 7 9

Mantel, N. 1970. A technique of nonparametric multivariate analysis. *Biometrics* 26: 547–558. [3.4.1, 4.1, 4.4]

7

Mardia, K. V. 1971. Effect of nonnormality of some multivariate tests and robustness to nonnormality in linear model. *Biometrika* 58: 105–121. [4.11]

2 7 8

Margalef, R. 1958a. Information theory in ecology. *Gen. Syst.* 3: 36–71. [3.5.2]

2 5 7 8

Margalef, R. 1958b. Temporal succession and spatial heterogeneity in phytoplankton. In *Perspectives on marine biology*, A. A. Buzzati-Traverso, (Ed.), pp. 323–349. University of California Press, Berkeley, Calif. [3.5.2, 3.11]

1 2 7

Marriott, F. H. C. 1974. *The interpretation of multiple observations*. Academic Press, London. [2.3.9, 3.4.2, 3.4.3, 3.5.3, 3.7, 4.1, 4.4]

2 5

McErlean, A. J., J. A. Mihursky, and H. J. Brinkley. 1969. Determination of upper temperature tolerance triangles for aquatic organisms. *Chesapeake Sci*. **10**: 293–296. [4.2]

5 8

McGowan, J. A., and V. J. Fraundorf. 1966. The relationship between size of net used and estimates of zooplankton diversity. *Limnol. Oceanogr*. **11**: 456–469. [2.3.6, 3.5.2]

1 7

McIntosh, R. P. 1973. Matrix and plexus techniques. In *Ordination and classification of communities*, Part V, *Handbook of vegetation science*, R. H. Whittaker, (Ed.), pp. 159–191. W. Junk, The Hague. [3.11]

8

McIntosh, R. P. 1967. An index of diversity: The relation of certain concepts to diversity. *Ecology* **48**: 392–403. [3.5.2]

5 7

McLeese, D. W. 1956. Effects of temperature, salinity and oxygen on survival of lobsters. *J. Fish. Res. Bd. Can*. **13**: 247–272. [4.2]

2

Mead, R. 1974. A test for spatial pattern at several scales using data from a grid of continuous quadrats. *Biometrics* **30**: 295–307. [2.3.7]

1 7

Mead, R., and D. J. Pike. 1975. A review of response surface methodology from a biometric viewpoint. *Biometrics* **31**: 803–851. [3.11]

8

Menge, B. A., and J. P. Sutherland. 1976. Species-diversity gradients– Synthesis of roles of predation, competition, and temporal heterogeneity. *Am. Nat*. **110**: 351–369. [3.5.2]

4

Merlini, M. 1967. The freshwater clam as a biological indicator of radio-manganese (*Unio mancus*). In *Proceedings of the international symposium on radioecological concentration processes*, Stockholm, Sweden. p. 977. Pergammon Press, London. [3.7]

6

Meslow, E. C., and L. B., Keith. 1971. A correlation analysis of weather versus snowshoe hare population parameters. *J. Wild. Mgmt*. **35**(1): 1–15. [2.1.9]

5

Miller, G. E. 1972. Mercury concentrations in museum specimens of tuna and swordfish. *Science* **175**: 1121–1122. [3.7, 4.5]

1 5

Mills, E. L. 1969. The community concept in marine zoology, with comments on continua and instability in some marine communities. A review. *J. Fish. Res. Bd. Can*. **26**: 1415–1428. [4.4]

4

Minns, C. K. 1977. Analysis of pulp and paper mill effluent plume. *J. Fish. Res. Bd. Can*. **34**: 776–783. [2.3.7]

1 7

Moore, D. H. 1973. Evaluation of five discrimination procedures for binary variables. *J. Am. Stat. Assoc*. **68**: 399–404. [3.4.3]

4

Moore, J. E., and R. J. Love. 1977. Effect of pulp and paper mill effluent on the productivity of periphyton and phytoplankton. *J. Fish. Res. Bd. can.* **34**: 856–862. [3.7]

1 8

Moore, P. D. 1975. Changes in species diversity. *Nature* **254**: 104–105. [3.5.2]

2 6

Morris, R. F. 1954. A sequential sampling technique for spruce budworm egg surveys. *Can. J. Zool.* **32**: 302–313. [2.3.9]

7

Moss, W. W. 1967. Some new analytical and graphic approaches to numerical taxonomy with an example from the *Dermanyssidae* (Acari). *Syst. Zool.* **16**: 177–207. [3.11]

1 2 9

Mosteller, F., and R. E. K. Rourke. 1973. *Sturdy statistics: Nonparametric and order statistics*. Addison-Wesley, Reading, Mass. [2.1.7, 3.4.2]

2 6

Munn, R. E. 1970. *Biometerological methods*. Academic Press, New York. [3.8, 3.11, 4.2]

3 7

Nie, N. H., C. H. Hull, J. G. Jenkins, K. Steinbrenner, and D. H. Bent. 1975. *SPSS: Statistical package for the social sciences*. McGraw-Hill, New York. [3.4.3, 3.6, 3.10, 3.11]

1 2 7

Norris, J. M. 1971. Singular matrices in multiple discriminant analysis and classification procedures. *Pedobiologia* **11**: 410–416. [3.5.3]

3 6 7

Norris, J. M., and J. P. Barkham. 1970. A comparison of some Cotswold beechwoods using multiple-discriminant analysis. *J. Ecol.* **58**: 603–619. [3.4.3]

2 6 7

Noy-Meir, I. 1971. Multivariate analysis of desert vegetation. II. Qualitative/quantitative partition of heterogeneity. *Isr. J. Bot* **20**: 203–13. [3.10]

2 6 7

Noy-Meir, I. 1974. Catenation: Quantitative methods for the definition of coenoclines. *Vegetation* **29**: 89–99. [4.4]

7 9

Noy-Meir, I., and M. P. Austin. 1970. Principal component ordination and simulated vegetational data. *Ecology* **51**: 551–552. [4.4]

1 7

Noy-Meir, I., D. Walker, and W. T. Williams. 1975. Data transformations in ecological ordination. II. On the meaning of data standardization. *J. Ecol.* **63**: 778–800. [2.3.9, 4.4]

6 7

Noy-Meir, I., N. H. Tadmor, and G. Orshan. 1970. Multivariate analysis of desert vegetation. I. Association analysis at various quadrat sizes. *Isr. J. Bot.* **19**: 561–591. [3.4.1, 3.4.3]

2 4

Oakland, G. B. 1950. An application of sequential analysis to whitefish sampling. *Biometrics* **6**: 59–67. [2.3.9]

4

Obeng-Asamoa, E. K., and B. C. Parker. 1972. Seasonal changes in phytoplankton and water chemistry of Mountain Lake, Virginia. *Trans. Am. Micros. Soc.* **91**: 363–380. [4.2]

3

O'Leary, M., R. H. Lippert, and O. T. Spitz. 1966. FORTRAN IV and MAP program for computation and plotting of trend surfaces degrees 1 through 6. *Computer Contr.* 3, Kansas State Geological Survey, Lawrence, Kansas. [3.10, 3.11]

1 2 7

Olson, C. L. 1976. On choosing a test statistic in multivariate analysis of variance. *Psych. Bull.* **83**: 579–586. [3.6, 4.1]

1 2 7

Orloci, L. 1966. Geometric models in ecology. I. The theory and application of some ordination methods. *J. Ecol.* **54**: 193–215. [2.3.9, 4.4]

1 7

Orloci, L. 1967. Data centering: A review and evaluation with reference to component analysis. *Syst. Zool.* **16**: 208–212. [2.3.9, 4.4]

2 7

Orloci, L. 1968. Information analysis in phytosociology: Partition, classification and prediction. *J. Theoret. Biol.* **20**: 271–284. [4.1, 4.4]

1 2 7

Orloci, L. 1972. On objective functions of phytosociological resemblance. *Am. Midl. Nat.* **88**: 28–55. [3.4.1, 3.9, 4.3]

2 7

Orloci, L. 1973a. Ranking characters by a dispersion criterion. *Nature, Lond.* **244**: 371–373. [3.6, 4.1]

1 7

Orloci, L. 1973b. Ordination by resemblance matrices. In *Handbook of vegetation science*, Part V, *Ordination and classification of vegetation*, R. H. Whittaker (Ed.), pp. 251–286. W. Junk, The Hague. [4.4]

7

Orloci, L. 1974. On information flow in ordination. *Vegetatio* **29**: 11–16. [4.4]

1 3 7

Orloci, L. 1975a. *Multivariate analysis in vegetation research.* W. Junk, The Hague. [3.4.1, 3.6, 3.10, 4.1, 4.4]

1 7

Orloci, L. 1975b. Partition of information: Some formulae revisited. *Aust. J. Bot.* **23**: 977–979. [3.4.3]

2 7

Orloci, L. 1975c. Measurement of redundancy in species collections. *Vegetatio* **31**: 65–67. [3.6]

2 3 7

Orloci, L. 1976. Ranking species by an information criterion. *J. Ecol.* **64**: 417–419. [3.6, 3.10, 4.1]

2 7

Orloci, L., and M.-M. Mukkattu. 1973. The effect of species number and type of data on the resemblance structure of a phytosociological collection. *J. Ecol.* **61**: 37–46. [3.4.3, 3.6]

2

Otis, D. L., K. P. Burnham, G. C. White, and D. R. Anderson. In press. Statistical inference from capture data on closed animal populations. *Ecol. Monogr.* [3.7]

8

Paine, R. T. 1966. Food web complexity and species diversity. *Am. Nat.* **100**: 65–76. [3.5.2]

8

Paine, R. T. 1969. A note on trophic complexity and community stability. *Am. Nat.* **103**: 91–93. [3.5.2]

2

Paloheimo, J. 1963. Estimation of catchabilities and population sizes of lobsters. *J. Fish. Res. Bd. Can.* **20**: 59–88. [3.7]

5

Panella, G., C. MacClintock, and M. N. Thompson. 1968. Paleontological evidence of variations in length of synodic month since late Cambrian. *Science* **162**: 792–796. [4.5]

2 7 9

Parker, R. A. 1968. Simulation of an aquatic ecosystem. *Biometrics* **24**: 803–821. [2.1.7]

4

Patalas, K. 1971. Crustacean plankton communities in forty-five lakes in the Experimental Lakes Area, northwestern Ontario. *J. Fish. Res. Bd. Can.* **28**: 231–244. [4.4]

2

Paulik, G. J. 1963. Estimates of mortality rates from tag recoveries. *Biometrics* **19**: 28–57. [3.7]

8

Peet, R. K. 1975. Relative diversity indices. *Ecology* **56**: 496–498. [3.3.2, 3.5.3]

5

Peterson, C. H. 1976. Relative abundances of living and dead molluscs in two California lagoons. *Lethaia* **9**: 137–148. [3.4.1]

5

Phillips, D. J. H. 1976. Common mussel *Mytilus edulis* as an indicator of pollution by zinc, cadmium, lead and copper. 2. Relationship of metals in mussel to those discharged by industry. *Mar. Biol.* **38**: 71–80. [3.7]

1 8

Pianka, E. R. 1966. Latitudinal gradients in species diversity: A review of concepts. *Am. Nat.* **100**: 33–46. [3.5.2]

1 2 7 8

Pielou, E. C. 1969. *An introduction to mathematical ecology.* Wiley, New York. [2.3.6, 3.4.1, 3.4.3, 3.5.2, 3.6, 4.1, 4.4]

2 7 8 9

Pielou, E. C. 1972. Measurement of structure in animal communities. In *Ecosystem structure and function,* Proceedings of the 31st Annual Biology Colloquium, John A. Wiens (Ed.). Oregon State University Press. [3.4.1, 3.5.2]

7

Pielou, E. C. 1974. Vegetation zones: Repetition of zones on a monotonic environmental gradient. *J. Theor. Biol.* **47**: 485–489. [2.1.9]

2 7

Pielou, E. C. 1975. Ecological models on an environmental gradient. In *Applied statistics,* R. P. Gupta (Ed.). North-Holland Publ. Co. [2.1.9, 3.4.1]

2 7

Pielou, D. P., and E. C. Pielou. 1967. The detection of different degrees of coexistence. *J. Theoret. Biol.* **16**: 427–437. [3.4.1]

2

Pikul, R. 1974. Development of environmental indices. In *Statistical and mathematical aspects of pollution problems*, J. W. Pratt (Ed.). Marcel Dekker, New York. [3.5.1]

8

Pimentel, D. 1961. Species diversity and insect population outbreaks. *Ann. Entomol. Soc. Am.* **54**: 76–86. [3.5.2]

1 2 3 7

Pimentel, R. A. 1978. *Morphometrics: The multivariate analysis of biological data*. Kendall/Hunt, Dubuque, Iowa. [3.4.3, 4.4]

1 2 7 8 9

Poole, R. W. 1974. *An introduction to quantitative ecology*. McGraw-Hill, New York. [1.2, 2.3.9, 3.5.2, 3.7, 4.2, 4.4]

1 3 7

Press, S. J. 1972. *Applied multivariate analysis*. Holt, Rinehart and Winston, New York. [3.4.2, 3.10]

1 7

Quenouille, M. H. 1957. *The analysis of multiple time series*. Hafner, New York. [4.2]

1 9

Raeside, D. E. 1976. Monte carlo principles and applications. *Phys. Med. Biol.* **21**: 181–197. [2.1.7]

2 7

Rao, C. R. 1966. Covariance adjustment and related problems in multivariate analysis. In *Multivariate analysis*, P. R. Krishnaiah (Ed.). Academic Press, New York. [3.6]

1 7

Rao, C. R. 1972. Recent trends of research work in multivariate analysis. *Biometrics* **28**: 3–22. [3.5.3, 3.7]

6 8

Recher, R. F. 1969. Bird species diversity and habitat diversity in Australia and North America. *Am. Nat.* **103**: 75–80. [3.5.2]

1

Redfield, A. C. 1958. The inadequacy of experiment in marine biology. In *Perspectives in marine biology*, A. A. Buzzati-Traverso (Ed.), pp. 17–26. University of California Press, Berkeley, Calif. [2.1.2, 2.1.4]

5

Reish, D. J. 1959. A discussion of the importance of screen size in washing quantitative marine bottom samples. *Ecology* **40**: 307–309. [2.3.6]

2 5

Reys, J.-P. 1971. Analyses statistiques de la microdistribution des espèces benthiques de la région de Marseille. *Tethys* **3**: 381–403. [2.3.9]

5 7

Reys, J.-P. 1976. Les peuplements benthiques (zoobenthos) de la région de Marseille (France): Aspects méthodologiques de la délimitation des peuplements par les méthodes mathématiques. *Mar. Biol.* **36**: 123–134. [4.4]

4 5

Rhoads, D. C., and G. Panella. 1970. The use of molluscan shell growth patterns in ecology and paleoecology. *Lethaia* **3**: 143–161. [3.7, 4.5]

2

Richards, F. J. 1959. A flexible growth functions for empirical use. *J. Exp. Bot.* **10**: 290–300. [3.7]

1 2

Ricker, W. E. 1958. Handbook of computations for biological statistics of fish populations. *Fish. Res. Bd. Can. Bull.* **119**, Queens Printer, Ottawa. [3.7]

6 7

Riechert, S. E. 1976. Web-site selection in the desert spider *Agelenopsis aperta. Oikos* **27**: 311–315. [3.6]

2

Robson, D. S., K. H. Pollack, and D. L. Solomon. 1972. *Test for mortality and recruitment in a k-sample tag-recapture experiment.* Biometrics Unit Publication BU-422-M, Cornell University, Ithaca, N.Y. [3.7]

2

Rohde, C. A. 1976. Composite sampling. *Biometrics* **32**: 273–282. [2.3.2]

3 7

Rohlf, F. J. 1969. GRAFPAC, graphic output subroutines for the GE635 Computer. *State Geol. Surv. Comput. Contrib.* No. 36. Lawrence, Kansas. 50 pp. [3.10, 3.11]

3 7

Rohlf, F. J. 1970. Adaptive hierarchical clustering schemes. *Syst. Zool.* **19**: 58–82. [3.4.2, 3.9, 3.11, 4.4]

1 7

Rohlf, F. J. 1974. Methods of comparing classifications. *Ann. Rev. Ecol. Syst.* **5**: 101–113. [4.4]

2 7

Rohlf, F. J. 1975. Generalization of gap test for detection of multivariate outliers. *Biometrics* **31**: 93–101. [3.9]

3 7

Rohlf, F. J., J. Kishpaugh, and D. Kirk. 1974. *NT-SYS, Numerical taxonomy system of multivariate statistical programs.* SUNY, Stony Brook, N. Y. [3.4.2, 3.4.3, 3.9, 3.10]

5 8

Rosenberg, R. 1971. Recovery of the littoral fauna in Saltkällefjord subsequent to discontinued operations of a sulphite pulp mill. *Thalassia Jugosl.* **7**: 341–351. [3.5.2]

5 8

Rosenberg, R. 1972. Benthic faunal recovery in a Swedish fjord following closure of a sulphite pulp mill. *Oikos* **23**: 92–108. [3.5.2]

5 8

Rosenberg, R. 1973. Succession in benthic macrofauna in a Swedish fjord subsequent to the closure of a sulphite pulp mill. *Oikos* **24**: 1–16. [2.1.8, 3.5.2]

5

Rounsefell, G. A. 1963. Marking fish and invertebrates. Fishery Leaflet 549, U.S. Fish and Wildlife Service. [3.7]

5

Rounsefell, G. A., and A. Dragovich. 1966. Correlation between oceanographic factors and abundance of the Florida red-tide (*Gymnodinium breve* Davis), 1954–61. *Bull. Mar. Sci.* **16**: 404–422. [2.3.9, 3.4.2]

5

Russell, H. J. 1972. Use of a commercial dredge to estimate a hardshell clam population by stratified random sampling. *J. Fish. Res. Bd. Can.* **29**: 1731–1735. [2.3.7]

4 8

Sager, P. E. and A. D. Hasler. 1969. Species diversity in lacustrine phytoplankton. I. The components of the index of diversity from Shannon's formula. *Am. Nat.* **103**: 51–59. [3.5.2]

5

Saila, S. B., R. A. Pikanowski, and D. S. Vaughan. 1976. Optimum strategies for sampling benthos in the New York Bight. *Estuarine Coastal Mar. Sci.* **4**: 119–128. [3.6]

4

Saila, S. B., T. T. Polgar, D. J. Sheeky, and J. M. Flowers. 1972. Correlations between alewife activity and environmental variables at a fishway. *Trans. Am. Fish. Soc.* **101**: 583–594. [3.7, 4.2]

1 2

Sampford, M. R. 1962. *An introduction to sampling theory.* Oliver & Boyd, Edinburgh. [2.3.5, 2.3.8]

5 8

Sanders, H. L. 1968. Marine benthic diversity: A comparative study. *Am. Nat.* **102**: 243–282. [3.5.2]

5 8

Sanders, H. L. 1969. Benthic marine diversity and the stability-time hypothesis. In *Diversity and stability in ecological systems*, G. M. Woodwell and H. H. Smith (Eds.). Brookhaven Symposium in Biology No. 22. [3.5.2]

4 7

Sandilands, R. G. 1977. Effect of pulp mill effluent on the surficial sediments of western Nipigon Bay, Lake Superior. *J. Fish. Res. Bd. Can.* **34**: 817–823. [4.4]

6

Schanda, E. (Ed.). 1976. Remote sensing for environmental sciences. Springer-Verlag, New York. [4.3]

2 7 9

Scott, A. J., and M. Knott. 1974. A cluster analysis method for grouping means in the analysis of variance. *Biometrics* **30**: 507–512. [4.4]

2

Seber, G. A. F. 1965. A note on the multiple-recapture census. *Biometrika* **52**: 249–259. [3.7]

1 2

Seber, G. A. F. 1973. *The estimation of animal abundance.* Griffin. [3.7]

5

Sellmer, G. P. 1956. A method for the separation of small bivalve molluscs from sediments. *Ecology* **37**: 206. [2.3.6]

3 7 9

Service, J. 1972. *A user's guide to the statistical analysis system* (SAS designed and implemented by A. J. Barr and J. H. Goodnight, Department of Statistics, North Carolina State University). North Carolina State University, Raleigh, N.C. [2.1.7, 3.4.2, 3.4.3, 3.6, 3.9, 3.10, 3.11, 4.1, 4.2,]

1 8

Shannon, C. E., and W. Weaver. 1949. *The mathematical theory of communication.* University of Illinois Press, Urbana. [3.5.2]

4 8

Sheldon, A. L. 1968. Species diversity and longitudinal succession in stream fishes. *Ecology* **49**: 193–197. [3.5.2]

8

Sheldon, A. L. 1969. Equitability indices: Dependence on the species count. *Ecology* **50**: 466–467. [3.5.2]

3

Siccama, T. G. 1972. A computer technique for illustrating three variables in a pictogram. *Ecology* **53**: 177–181. [3.11]

1 2

Siegel, S. 1956. *Nonparametric statistics for the behavioral sciences*. McGraw-Hill, New York. [2.1.7, 3.4.1, 3.4.2, 4.2]

8

Simberloff, D. 1972. Properties of the rarefaction diversity measurement. *Am. Nat.* **106**: 414–418. [3.5.2]

8

Simpson, E. H. 1949. Measurement of diversity. *Nature* **163**: 688. [3.5.2]

1

Skellam, J. G. 1969. Models, inference, and strategy. *Biometrika* **25**: 457–475. [2.1.1, 2.1.2, 2.1.4]

1

Slobodkin, L. B. 1968. Aspects of the future of ecology. *Biosci.* **18**: 16–23. [3.7, 4.2]

2 3

Slocomb, J., B. Stauffer, and K. L. Dickson. 1977. On fitting the truncated lognormal distribution to species-abundance data using maximum likelihood estimation. *Ecology* **58**: 693–696. [3.10]

3 6 7

Smartt, P. F. M., S. E. Meacock, and J. M. Lambert. 1974. Investigations into the properties of quantitative vegetational data. I. Pilot study. *J. Ecol.* **62**: 735–760. [3.4.1]

4

Smith, A. L., R. H. Green, and A. Lutz. 1975. Uptake of mercury by freshwater clams (Family Unionidae). *J. Fish. Res. Bd. Can.* **32**: 1297–1303. [3.7]

1 3

Smith, B. T., J. M. Boyle, B. S. Garlow, Y. Ikebe, V. C. Klema, and C. B. Moler. 1976. *Matrix eigensystem routines, EISPACK guide*. Springer-Verlag, New York. [3.10]

1 7

Smith, R. W. 1976. Numerical analysis of ecological survey data. Ph.D. thesis, University of Southern California, Los Angeles. [3.11]

5 7 8

Smith, R. W., and C. S. Greene. 1976. Biological communities near submarine outfall. *J. Water Pollut. Control Fed.* **48**: 1894–1912. [4.4]

1 8

Smith, W., and J. F. Grassle. 1977. Sampling properties of a family of diversity measures. *Biometrics* **33**: 283– [3.5.2]

1 7

Sneath, P. H. A. 1967. Some statistical problems in numerical taxonomy. *Statistician* **17**: 1–12. [4.4]

1 2

Snedecor, G. W., and W. G. Cochran. 1967. *Statistical methods*, 6th ed. Iowa State University Press, Ames. [3.5.3]

1 2 7

Sokal, R. R., and P. H. A. Sneath. 1963. *Principles of numerical taxonomy*. Freeman, San Francisco. [3.6]

1 2

Sokal, R. R., and R. J. Rohlf. 1969. *Biometry*. Freeman, San Francisco. [2.1.1, 2.3.9, 3.5.3, 4.3]

1 2

Sokal, R. R., and F. J. Rohlf. 1973. *Introduction to biostatistics*. Freeman, San Francisco. [2.1.1, 2.3.2, 2.3.9, 3.5, 3.5.1, 3.5.3]

1 2 4 5 6 7 8

Southwood, T. R. E. 1966. *Ecological methods*. Methuen, London. [2.3.9, 3.7, 4.2]

4

Spence, J. A., and H. B. N. Hynes. 1971. Differences in benthos upstream and downstream of an impoundment. *J. Fish. Res. Bd. Can.* **28**: 35–43. [3.6]

5 7

Spight, T. M. 1976. Just another benthic study. *Ecology* **57**: 622–623. [2.3.1, 3.6]

2 4 5

Sprague, J. B. 1969. Measurement of pollutant toxicity to fish. I. Bioassay methods for acute toxicity. *Water Res.* **3**: 793–821. [4.2]

4 7

Sprules, W. G. 1977. Crustacean zooplankton communities as indicators of limnological conditions: An approach using principal component analysis. *J. Fish. Res. Bd. Can.* **34**: 962–975. [3.4.2, 4.4]

4

Stauffer, J. R., Jr., K. L. Dickson, and J. Cairns, Jr. 1974. A field evaluation of the effects of heated discharges on fish distribution. *Water Res. Bull.* **10**: 860–876. [3.7]

1 2

Steel, R. G. D., and J. H. Torrie. 1960. *Principles and procedures of statistics with special reference to the biological sciences*. McGraw-Hill, New York. [2.1.5, 2.3.3, 2.3.7, 2.3.9, 3.5.3]

5 7

Stephenson, W., Y. I. Raphael, and S. D. Cook. 1976. The macrobenthos of Bramble Bay, Moreton Bay, Queensland. *Mem. Queensl. Mus.* **17**: 425–447. [3.8]

5

Sturesson, U., and R. A. Reyment. 1971. Some minor chemical constituents of the shell of *Macoma balthica*. *Oikos* **22**: 414–416. [3.7]

2

Suits, D. B. 1957. Use of dummy variables in regression equations. *J. Am. Stat. Assoc.* **52**: 548–551. [3.4.3]

5

Swan, E. F. 1961. Some uses of colored materials in marine biological research. *Turtox News* **39**: 290–293. [3.7]

1 7 9

Swan, J. M. A. 1970. An examination of some ordination problems by use of simulated vegetational data. *Ecology* **51**: 89–102. [4.4]

2

Taylor, C. C. 1953. Nature of variability in trawl catches. *Fish. Bull.* **83**, U.S. Fisheries and Wildlife Ser. 54. [2.3.9]

2

Taylor, L. R. 1961. Agregation, variance and the mean. *Nature* **189**: 732–735. [2.3.9, 3.8]

5

Tevesz, M. J. S. 1972. Implications of absolute age and season of death data compiled for recent *Gemma gemma*. *Lethaia* **5**: 31–38. [3.7]

1 8

Thomas, W. A., G. Goldstein, and W. H. Wilcox. 1973. *Biological indicators of environmental quality. A bibliography of abstracts*. Ann Arbor Science, Ann Arbor, Mich. [3.5.2]

1 7

Thorpe, R. S. 1976. Biometric analysis of geographic variation and racial affinities. *Biol. Rev.* **51**: 407–452. [3.4.3, 3.6, 3.11]

5 7 8

Tietjen, J. H. 1976. Distribution and species diversity of deep-sea nematodes off North Carolina. *Deep-Sea Res.* **23**: 755–768. [3.5.2]

5 7

Tietjen, J. H. 1971. Ecology and distribution of deep-sea meiobenthos off North Carolina. *Deep-Sea Res.* **18**: 941–957. [3.11]

5

Tonolli, V., and L. Tonolli. 1958. Irregularities of distribution of plankton communities: considerations and methods. In *Perspectives in marine biology*, A. A. Buzzati-Traverso (Ed.) University of California Press, Berkeley, Calif. [2.3.7]

6 8

Tramer, E. J. 1969. Bird species diversity: Component of Shannon's formula. *Ecology* **50**: 927–929. [3.5.2]

4 8

Tramer, E. J., and P. M. Rogers. 1973. Diversity and longitudinal zonation in fish populations of two streams entering a metropolitan area. *Am. Midl. Nat.* **90**: 366–374. [3.5.2]

1

Usher, M. B. 1972. Developments in the Leslie Matrix model. In *Mathematical models in ecology*, J. N. R. Jeffers (ed.), pp. 29–60. Blackwell, Oxford. [4.2]

4 7

Vascotto, G. L. 1976. The zoobenthic assemblage of four central Canadian lakes and their potential use as environmental indicators. Ph.D. Thesis, University of Manitoba, Winnipeg, Canada. [3.6]

2

Wald, A., and I. Wolfowitz. 1943. An exact test for randomness in the nonparametric case based on serial correlations. *Ann. Math. Stat.* **14**: 378–388. [4.2]

1 5

Waldichuk, M. 1973. Trends in methodology for evaluation of effects of pollutants on marine organisms and ecosystems. *Crit. Rev. Environ. Control* **3**: 167–211. [4.2]

Watson, J., and F. Crick. 1953. Molecular structure of nucleic acids: A structure for deoxyribose nucleic acid. *Nature* **171**: 737–738. [2.1.8]

1

Watt, K. E. F. 1968. *Ecology and resource management*. McGraw-Hill, New York. [3.6, 3.7]

6

Weiss, H. 1971. Mercury in a Greenland ice sheet: Evidence of recent input by man. *Science* 174: 692–694. [4.5]

3 7

Wermuth, N., T. Wehner, and H. Gönner. 1976. Finding condensed descriptions for multi-dimensional data. *Comput. Programs Biomed.* 6: 23–38. [3.6, 3.10]

6 7

West, N. E. 1966. Matrix cluster analysis of montane forest vegetation of the Oregon Cascades. *Ecology* 47: 975–979. [3.5.1]

4

Westlake, G. F., and W. H. van der Schalie. 1977. Evaluation of an automated biological monitoring system at an industrial site. In *Biological monitoring of water and effluent quality*, ASTM STP607, J. Cairns, Jr., K. L. Dickson, and G. F. Westlake (Eds.), pp. 30–37. American Society for Testing and Materials, Philadelphia. [3.7]

3

White, E. G. 1971. A versatile Fortran computer program for the capture-recapture stochastic model of G. M. Jolly. *J. Fish. Res. Bd. Can.* 28: 443–445. [3.7, 3.10]

1 7

Whittaker, R. H. 1962. Classification of natural communities. *Bot. Rev.* 28: 1–239. [4.4]

1

Whittington, H. B., and C. P. Hughes. 1972. Ordovician geography and faunal provinces deduced from trilobite distribution. *Phi. Trans. Roy. Soc. Lond. B.* 263: 235–278. [3.4.2, 3.11]

4

Whittle, D. M., and K. W. Flood. 1977. Assessment of the acute toxicity, growth impairment, and flesh tainting potential of a bleached kraft mill effluent on rainbow trout (*Salmo gairdneri*). *J. Fish. Res. Bd. Can.* 34: 869–878. [3.7, 4.2]

2

Wilcoxon, F., and R. A. Wilcox. 1964. *Some rapid approximate statistical procedures.* Lederle Laboratories, Pearl River, N.Y. [2.3.9]

4 8

Wilhm, J. L., and T. C. Dorris. 1968. Biological parameters for water quality criteria. *Biosci.* 18: 477–481. [3.5.2, 4.3]

3 7

Wilkinson, L., and L. R. Huesmann. 1973. Use of APL in teaching multivariate data analysis. *Behav. Res. Methods Instrum.* 5: 209–211. [3.10]

2 7

Wilks, S. S. 1963. Multivariate statistical outliers. *Sankhya Ser. A* 25: 407–426. [3.9]

4

Williams, D. D., and H. B. Hynes. 1974. The occurence of benthos deep in the substratum of a stream. *Freshwater Biol.* 4: 233–256. [2.3.6]

1 7

Williams, W. T. 1971. Principles of clustering. *Ann. Rev. Ecol. Syst.* 2: 303–326. [4.4]

2 7

Williams, W. T., and M. B. Dale. 1962. Partition correlation matrices for heterogeneous quantitative data. *Nature, Lond.* 196: 602. [3.4.3]

2 6 7

Williams, W. T., and J. M. Lambert. 1959. Multivariate methods in plant ecology. I. Association-analysis in plant communities. *J. Ecol.* **47**: 83–101. [3.4.1, 3.6, 4.2, 4.4]

2 3 6 7

Williams, W. T., and J. M. Lambert. 1960. Multivariate methods in plant ecology. II. The use of an electronic digital computer for association-analysis. *J. Ecol.* **48**: 689–710. [3.4.1, 3.6, 4.2, 4.4]

7

Williams, W. T., and G. N. Lance. 1965. Logic of computer-based intrinsic classifications. *Nature* **207**: 159–161. [2.1.2, 4.4]

1 7

Williams, W. T., and G. N. Lance. 1968. Choice of strategy in the analysis of complex data. *Statistician* **18**: 31–43. [4.4]

2 5 7

Williams, W. T., and W. Stephenson. 1973. The analysis of three-dimensional data (sites *x* species *x* times) in marine ecology. *J. Exp. Mar. Biol. Ecol.* **11**: 207–227. [2.3.9, 4.1]

1 2 6 7 8

Williams, W. T., G. N. Lance, L. J. Webb, J. G. Tracey, and M. B. Dale. 1969. Studies in the numerical analysis of complex rain-forest communities. III. The analysis of successional data. *J. Ecol.* **57**: 515–535. [3.5.2, 4.2]

1 2 4 5

Winberg, G. G. 1971. *Methods for the estimation of production of aquatic animals.* Translated by A. Duncan. Academic Press, London. [3.7]

3 7

Wishart, D. 1975. *CLUSTAN IC user's manual.* University College, London. [3.4.3, 3.10, 3.11, 4.4]

2 7

Wolfe, J. H. 1970. Pattern clustering by multivariate mixture analysis. *Multivar. Behav. Res.* **5**: 329–350. [3.5.3, 3.9, 4.4]

7

Woodward, J. A., and J. E. Overall. 1976. Factor-analysis of rank-ordered data: An old approach revisited. *Psych. Bull.* **83**: 864–867. [3.4.2, 3.5.3]

5 6

Wurster, C. F., and D. W. Wingate. 1968. DDT residues and declining reproduction in the Bermuda petrel. *Science* **159**: 979–981. [3.8]

2

Zahl, S. 1974. Application of the *S*-method to the analysis of spatial pattern. *Biometrics* **30**: 513–524. [2.3.7]

5 8

Zimmerman, M. S., and R. J. Livingston. 1976. Effects of kraft-mill effluents on benthic macrophyte assemblage in a shallow-bay system. (Apalachee Bay, North Florida, U.S.A.). *Mar. Biol.* **34**: 297–312. [3.5.2]

INDEX